Milestones in Drug Therapy
MDT

Entry Inhibitors in HIV Therapy

Edited by Jacqueline D. Reeves and Cynthia A. Derdeyn

Birkhäuser Verlag
Basel · Boston · Berlin

Editors
Jacqueline D. Reeves
Monogram Biosciences
345 Oyster Point Blvd.
South San Francisco, CA 94080
USA

Cynthia A. Derdeyn
Department of Pathology and
Laboratory Medicine
Emory Vaccine Center
Emory University
954 Gatewood Rd, Suite 1024
Atlanta, GA 30329
USA

Library of Congress Control Number: 2007926128

Bibliographic information published by Die Deutsche Bibliothek
Die Deutsche Bibliothek lists this publication in the Deutsche Nationalbibliografie; detailed bibliographic data is available in the internet at http://dnb.ddb.de

ISBN 978-3-7643-7782-3 Birkhäuser Verlag AG, Basel - Boston - Berlin

Part of Springer Science+Business Media
Printed on acid-free paper produced from chlorine-free pulp. TFC
Cover illustration: Targeting the conformational changes critical to membrane fusion. With kind permission of Eric Hunter, Emory Vaccine Research Center, Atlanta.

Printed in Germany
ISBN 978-3-7643-7782-3 e-ISBN 978-3-7643-7783-0

9 8 7 6 5 4 3 2 1 www. birkhauser.ch

Contents

List of contributors

Louis Alexander, Department of Virology, Bristol-Myers Squibb Pharmaceutical Research Institute, 5 Research Parkway, Wallingford, CT 06492, USA

Kabamba B. Alexandre, AIDS Virus Research Unit, National Institute for Communicable Diseases, Private Bag X4, Sandringham 2131, Johannesburg, South Africa

James M. Binley, Torrey Pines Institute for Molecular Studies, 3550 General Atomics Court, San Diego, CA 92121, USA

Tonie Cilliers, AIDS Virus Research Unit, National Institute for Communicable Diseases, Private Bag X4, Sandringham 2131, Johannesburg, South Africa

Eoin Coakley, Monogram Biosciences, 345 Oyster Point Blvd, South San Francisco, CA 94080, USA; e-mail: ecoakley@monogrambio.com

Mia Coetzer, AIDS Virus Research Unit, National Institute for Communicable Diseases, Private Bag X4, Sandringham 2131, Johannesburg, South Africa; present address: The Scripps Research Institute, 10550 North Torrey Pines Road, La Jolla, CA 90237, USA

Robert W. Doms, Department of Microbiology, University of Pennsylvania, 225 Johnson Pavilion, 3610 Hamilton Walk, Philadelphia, PA 19104, USA; e-mail: doms@mail.med.upenn.edu

Tammy Evans-Strickfaden, Division of HIV/AIDS Prevention, Centers for Disease Control and Prevention, 1600 Clifton Rd (ms A25), Atlanta, GA 30333, USA

Elin S. Gray, AIDS Virus Research Unit, National Institute for Communicable Diseases, Private Bag X4, Sandringham 2131, Johannesburg, South Africa

Michael L. Greenberg, Trimeris Inc., 3500 Paramount Parkway, Morrisville, NC 27560, USA; e-mail: mgreenberg@trimeris.com

Roy M. Gulick, Weill Medical College of Cornell University, Cornell HIV Clinical Trials Unit, Box 566, 525 East 68th Street, New York 10021 USA; e-mail: rgulick@med.cornell.edu

Clyde Hart, Division of HIV/AIDS Prevention, Centers for Disease Control and Prevention, 1600 Clifton Rd (ms A25), Atlanta, GA 30333, USA; e-mail: ceh4@CDC.GOV

Sonya Heath, Department of Medicine, Division of Infectious Diseases, Rm 328B Community Care Bldg, 908 20th Street South, Birmingham, AL 35294-2050, USA; e-mail: heaths@uab.edu

John Kadow, Department of Chemistry, Bristol-Myers Squibb Pharmaceutical Research Institute, 5 Research Parkway, Wallingford, CT 06492, USA

J. Michael Kilby, UAB 1917 Clinic University of Alabama at Birmingham,

CCB Room 142, 908 20th Street South, Birmingham, AL 35294-2050, USA; e-mail: mkilby@uab.edu

Pin-fang Lin, Department of Virology, Bristol-Myers Squibb Pharmaceutical Research Institute, 5 Research Parkway, Wallingford, CT 06492, USA; e-mail: PinFang.Lin@bms.com

Donald E. Mosier, Department of Immunology, The Scripps Research Institute, 10550 North Torrey Pines Road, La Jolla, CA 92037, USA; e-mail: dmosier@scripps.edu

Penny L. Moore, AIDS Virus Research Unit, National Institute for Communicable Diseases, Private Bag X4, Sandringham 2131, Johannesburg, South Africa

Lynn Morris, AIDS Virus Research Unit, National Institute for Communicable Diseases, Private Bag X4, Sandringham 2131, Johannesburg, South Africa; e-mail: lynnm@nicd.ac.za

Martine Peeters, UMR145, Centre IRD de Montpellier, 911 av Agropolis BP64501, 34394 Montpellier Cedex1, France; e-mail: Martine.Peeters@mpl.ird.fr

Stefan Pöhlmann, Institute for Clinical and Molecular Virology and Nikolaus-Fiebiger-Center for Molecular Medicine, University Erlangen-Nürnberg, Schlossgarten 4, 91054 Erlangen, Germany; e-mail: snpoehlm@viro.med.uni-erlangen.de

Julie M. Stritzki, Department of Virology, Schering-Plough Research Institute, 2015 Galloping Hill Road, K15, E405C/4945, Kenilworth, NJ 07033, USA; e-mail: julie.stritzki@spcorp.com

John C. Tilton, Department of Microbiology, University of Pennsylvania, 225 Johnson Pavilion, 3610 Hamilton Walk, Philadelphia, PA 19104, USA

Michel J. Tremblay, Laboratory of Human Immuno-Retrovirology, Research Center in Infectious Diseases, RC709 CHUL Research Center, 2705 Laurier Blvd, Quebec, Canada, G1V 4G2, and Faculty of Medicine, Laval University, Quebec, Canada

Laurence Vergne†, UMR145, Centre IRD de Montpellier, 911 av Agropolis BP64501, 34394 Montpellier Cedex1, France

Wei Wang, Food and Drug Administration, Center for Biologics Evaluation and Research, Bldg. 29, Room 532, 29 Lincoln Drive, Bethesda, Maryland 20892-4555, USA

Carol D. Weiss, Food and Drug Administration, Center for Biologics Evaluation and Research, Bldg. 29, Room 532, 29 Lincoln Drive, Bethesda, Maryland 20892-4555, USA; e-mail: cdweiss@helix.nih.gov

Preface

The entry of HIV into cellular targets is mediated by the envelope protein (Env) which studs the viral surface. A major milestone for inhibition of HIV entry was achieved in 2003 with the approval of the HIV fusion inhibitor enfuvirtide. A number of other entry inhibitors are currently being developed with diverse mechanisms of action, including (i) interfering with relatively non-specific Env-cell surface attachment factor interactions, (ii) inhibiting specific receptor and coreceptor interactions, and (iii) blocking Env transition through conformational intermediate states. Major challenges facing entry inhibitor development include the extensive sequence diversity and remarkable plasticity of Env. Env diversity can give rise to marked variability in baseline susceptibility of HIV strains to entry inhibitors, as opposed to typically minor variation in baseline susceptibilities to HIV reverse transcriptase and protease inhibitors.

Entry Inhibitors in HIV Therapy presents the current status of this relatively new and highly dynamic class of inhibitors and provides a unique overview of obstacles and considerations for HIV entry inhibition compared to other antiretroviral targets. It will be of interest to research scientists as well as clinicians.

The introductory chapters of this book provide an overview of HIV entry, entry inhibition and envelope diversity. The first chapter, by Tilton and Doms, reviews current knowledge of how Env mediates entry and presents an overview of entry inhibitors. Vergne and Peeters then discuss the challenge of genetic diversity in the HIV envelope.

Subsequent chapters of this volume feature current information on individual classes of entry inhibitors that target each step of the virus entry pathway, from attachment to membrane fusion, with an emphasis on the complex determinants of entry inhibitor susceptibility, resistance mechanisms, and how these issues create new challenges for antiretroviral therapy. Pöhlmann and Tremblay review inhibitors that block HIV cell surface attachment and Lin, Kadow and Alexander discuss inhibitors that target Env interactions with CD4. Strizki and Mosier review inhibitors of Env-coreceptor interactions and Wang and Weiss describe inhibitors that target HIV fusion. Studies of entry inhibitors as microbicides are presented by Hart and Evans-Strickfaden and the use of entry inhibitors against non-subtype B viruses is discussed by Morris, Binley and colleagues. Coakley then discusses the clinical utility of coreceptor typing and entry inhibitor susceptibility testing.

The final chapters of this book highlight the clinical use of entry inhibitors and survey antiretroviral development. Heath and Kilby review the current status of entry inhibitors in clinical studies. The development and approval of enfuvirtide is detailed by Greenberg, then past and present drug development targets are discussed by Gulick.

In summary, this book presents a comprehensive and current overview of entry inhibitors from an expert panel of authors with diverse backgrounds and perspectives, incorporating many unrelenting successes against a backdrop of formidable challenges.

Jacqueline D. Reeves, San Francisco
Cynthia A. Derdeyn, Atlanta February 2007

Entry Inhibitors in HIV Therapy
Edited by Jacqueline D. Reeves and Cynthia A. Derdeyn

Introduction to entry inhibitors in the management of HIV infection

John C. Tilton and Robert W. Doms

Department of Microbiology, University of Pennsylvania, 225 Johnson Pavilion, 3610 Hamilton Walk, Philadelphia, PA 19104, USA

Introduction

The introduction of highly active antiretroviral therapy (HAART) has dramatically improved the survival of patients infected with human immunodeficiency virus (HIV). However, HAART is complicated by the continuing emergence of drug-resistant strains of HIV and toxicities associated with the antiretroviral agents [1, 2]. Furthermore, since the combination HAART regimens are incapable of eradicating HIV infection, lifelong therapy is required to avoid disease progression [3, 4]. Together, these factors necessitate the continual development of new antiretroviral agents that can be utilized against resistant viruses or that in combination with other agents can provide superior viral suppression with less toxicity.

While all stages of the HIV life cycle are potential targets for therapeutic intervention, HAART regimens have been predominantly focused on two viral enzymes, reverse transcriptase (RT) and protease. New antiretroviral agents under development include integrase inhibitors as well as compounds that target components of the viral entry pathway. These latter compounds are collectively known as entry inhibitors, and are the subject of this volume. Entry inhibitors are varied in that they can target three different steps in the viral entry pathway: CD4 binding (Chapter by Lin et al.), coreceptor binding (Chapter by Stritzki/Mosier), or membrane fusion (Chapters by Wang/Weiss and Greenberg). Regardless of their precise mechanism, all entry inhibitors target the viral envelope (Env) protein directly or, in the case of coreceptor inhibitors, indirectly. Thus, a major challenge to the clinical use of entry inhibitors is the impressive sequence diversity of the Env protein, which contributes to the significant variation in the baseline sensitivity of HIV isolates to these compounds (Chapter Vergne/Peeters). Patient-specific variation in host factors involved in the HIV entry process may also modulate the susceptibility of HIV to entry inhibitors and the development of resistance mutations. Resistance pathways to entry inhibitors are likely to be complex, and may alter viral tropism (and hence pathogenesis and disease course) by altering the manner in which Env interacts with host cell receptors. Together, these factors make entry inhibitors a particularly interesting class of antiretroviral agent that may shed significant light on HIV pathogenesis.

The viral Env protein

The Env protein of HIV-1 is the molecular determinant for viral attachment and membrane fusion. Env is synthesized as a single polypeptide precursor (gp160) that forms noncovalently associated homotrimers, and which is cleaved during transport to the cell surface into two subunits, gp120 and gp41. The gp120 subunit mediates receptor binding, while gp41 mediates the membrane fusion reaction. HIV-1 gp120 consists of five conserved (C1–C5) and five variable (V1–V5) domains [5], with the conserved domains contributing to the core of gp120, while the variable domains (and numerous *N*-linked glycosylation sites) are located near the surface of the molecule. The V1–V4 regions form exposed 'loops' anchored at their bases by disulfide bonds [6]

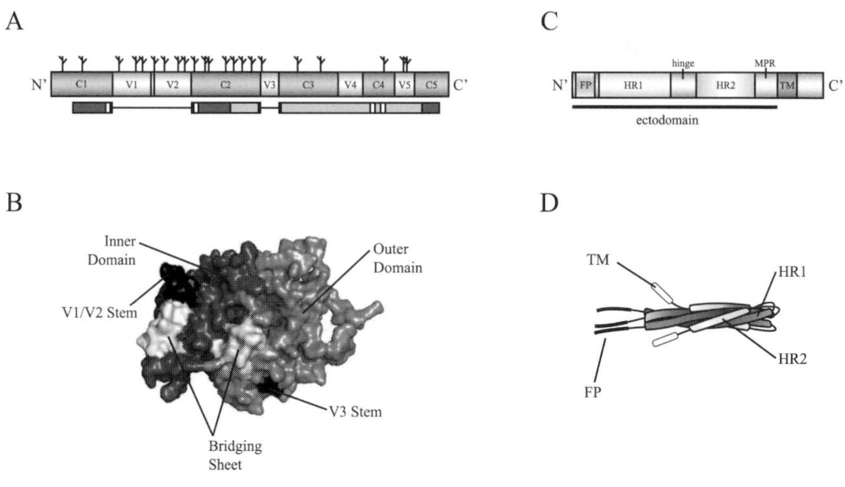

Figure 1. The HIV envelope proteins gp120 and gp41. (A) Schematic diagram of the HIV gp120 protein showing the five conserved (C1–C5) and five variable (V1–V5) domains. Positions of conserved glycosylation sites are indicated by branched chains on the diagram. The conserved "core" of gp120 with deletions of the V1/V2 and V3 loops, and truncations at the N' and C' termini, is depicted underneath the main gp120 molecule, and is shaded to match the domains indicated in the space-filling model of gp120. (B) Space-filling model of gp120 showing the major domains of the protein, including the inner domain (dark gray), outer domain (light gray), bridging sheets (white) and V1/V2 and V3 stems (black). (C) Schematic diagram of the HIV gp41 protein showing the fusion peptide (FP), heptad-repeat domains (HR1 and HR2), and the transmembrane anchor (TM). (D) A model of the gp41 protein in the post-fusion conformation, where HR1 and HR2 have interacted to form the six-helix bundle structure. Note the proximity of the fusion peptide (which has inserted into the host membrane) and the TM region (which is inserted into the viral membrane). The fusion inhibitor enfuvirtide acts by interfering with association of the HR1 and HR2 domains, blocking the formation of the six-helix bundle.

(Fig. 1A). The gp120 molecule has proven difficult to crystallize in its entirety, but several structures have been solved in recent years, including a deglycosylated HIV-1 gp120 bound to CD4 and lacking the V1–3 loops as well as containing truncations at the N and C termini [7], a similar molecule but containing the V3 loop [8], and a glycosylated form of SIV gp120 also lacking V1–3 and small portions of the N and C termini [9]. From these structures, it is evident that in its native state gp120 contains two distinct regions: an inner domain that is involved in interactions with gp41 and the formation of trimeric envelope spikes, and an outer domain that forms a large part of the exposed surface of the spikes and is heavily glycosylated. Binding of CD4 to gp120 induces significant conformational changes that result in the formation of a third domain termed the bridging sheet (Fig. 1B). This domain consists of two pairs of antiparallel β-sheets that link the inner and outer domains, and plays a major role in interacting with the viral coreceptors [10].

The gp41 protein consists of three distinct domains: an unusually large cytoplasmic domain on the inside of the viral membrane, a transmembrane (TM) anchor, and an ectodomain that extends from the surface of the virion. The ectodomain is the principal determinant of membrane fusion and contains a hydrophobic, N-terminal fusion peptide that is believed to insert into the cellular membrane and two heptad repeat (HR) sequences, HR1 and HR2, which are critical to the fusion process [11, 12] (Fig. 1C). The only approved member of the entry inhibitor class of antiretrovirals, enfuvirtide (Fuzeon, T20) acts by targeting the interaction of the two conserved HR domains [13] (Fig. 1D).

A major challenge in the design of entry inhibitors that target the viral Env protein is that the structure of the native, trimeric Env is not known. Recent electron tomography studies reveal the overall dimensions of Env trimers [14, 15], but more precise information will be needed to assist in structure-based drug design efforts.

The HIV-1 entry process

Entry of HIV-1 into cells involves three distinct stages: binding of gp120 to CD4, binding of gp120 to coreceptor, and gp41-mediated fusion of the viral and host membranes. The primary receptor for HIV-1 is CD4, a member of the immunoglobulin superfamily that is expressed on monocytes, macrophages, and subsets of dendritic cells. CD4 makes contact with the gp120 molecule at a depression near the intersection of the inner domain, outer domain, and bridging sheet [7] (Fig. 2A, B). CD4 binding actually appears to induce the formation of the bridging sheet domain itself, as the two pairs of β-sheets are spatially separated in a crystal structure of the unliganded core of SIV gp120 but come together to form a four-stranded sheet in the CD4-liganded conformation [7, 9] (Fig. 2C). Additional changes in gp120 occur with CD4 binding, including movement of the V1/V2 and V3 loop structures. As a result, CD4 binding not only induces the formation of the bridging sheet, it likely enhances

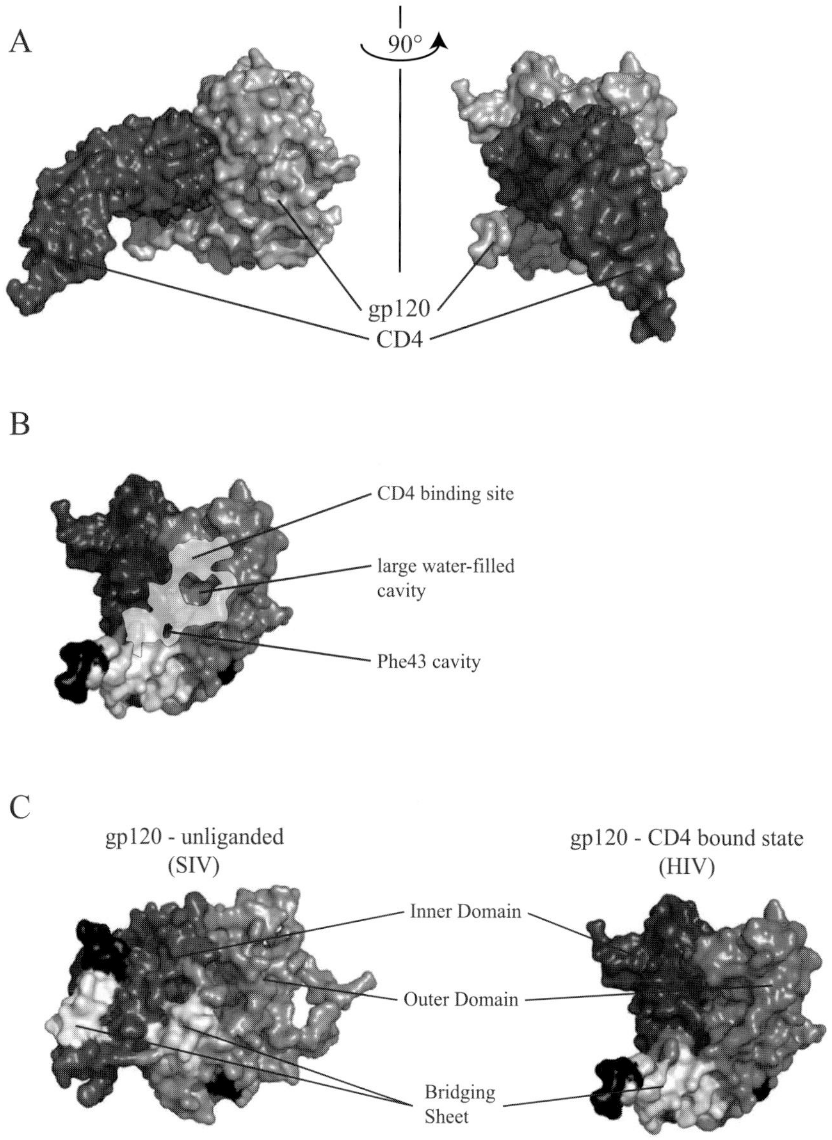

Figure 2. Space-filling models of gp120 interactions with CD4. (A) Models depicting the most distal immunoglobulin domain of the CD4 receptor (black) binding to CD4 gp120 (gray). (B) The CD4-binding site on the surface of gp120 (white highlight) overlaps the inner domain, outer domain, and the bridging sheet domain (white) of gp120. Two water-filled pockets, a large cavity and the Phe43 cavity, are conserved features of the CD4-gp120 interface and are thought to be important in the conformational changes in gp120 that accompany CD4 and coreceptor binding. (C) Interaction between gp120 and CD4 results in major conformational changes in gp120. The structure of gp120 in an unliganded form (left) shows the positions of the inner domain (dark gray), outer domain (light gray) and the β-sheets that comprise the bridging sheet (white). Note that in the unliganded state (left), the bridging sheet β-sheets are spatially separated, but come together to form a four-stranded domain in the CD4-bound conformation (right).

exposure of this region and orients it and the V3 loop towards the target cell membrane where they can engage the viral coreceptor [16, 17]. Unlike other regions of the gp120 molecule, the residues contacting CD4 are highly conserved and are devoid of carbohydrate [7]. These properties make the CD4-binding face of gp120 a logical target for small molecule inhibitors of gp120-CD4 binding.

In humans, the major coreceptors for HIV-1 are the chemokine receptors CCR5 and CXCR4, both of which are members of the seven-TM G protein-coupled receptor family [18–25]. These receptors are integral membrane proteins with a small extracellular pocket formed by three loops between TM segments. The N-terminal segment of the receptor also extends into the extracellular space. Both regions are involved in binding to gp120. The two main regions of gp120 that are involved in binding to coreceptor are the coreceptor-binding site formed by the bridging sheet and adjoining regions, and the V3 loop [26–29]. Several of the amino acids in the coreceptor-binding site are among the most highly conserved residues between HIV-1, HIV-2, and SIV [10, 30]. In contrast, the V3 loop is defined as one of the variable domains of gp120, but the length of the V3 loop is strictly conserved, with most HIV-1 isolates containing between 34 and 36 residues. V3 has a GPGR or GPGQ motif that forms a β-turn in the loop, a region that comprises the center of the 'tip' or 'crown' of V3. Binding of gp120 to the pocket of CCR5 appears to be dependent on the residues present at the 'crown' of the V3 loop [29]. Contact between residues in the tip of CCR5 and extracellular loop 2 have been shown to be particularly important for HIV entry [31–34]. These data are consistent with a recent crystal structure of gp120 in which the V3 loop is found to extend nearly 30 Å from its base towards the cellular membrane, where it could presumably make contact with the chemokine receptor pocket [8].

On the coreceptor molecules, the N terminus of CCR5 is rich in sulfated tyrosines and is highly acidic [35]. Mutagenic studies have indicated that these sulfotyrosines in the N-terminal extracellular region of CCR5 interact with gp120 by binding to conserved residues at the base of the V3 loop and may also make contact with the coreceptor binding site [28]. Indeed, sulfated peptides corresponding to this region inhibit infection by CCR5-tropic viruses [36, 37]. Binding of CXCR4 to gp120 appears to occur in a similar fashion [26, 38–41].

Binding of gp120 to coreceptor is believed to trigger further conformational changes in the envelope trimer that enable gp41 to mediate the fusion of viral and cellular membranes [42]. The structural rearrangements triggered by binding to CD4 and coreceptor are believed to allow the glycine-rich, hydrophobic fusion peptide at the N-terminal region of gp41 to insert into the target cell membrane. Following insertion of the fusion peptides, the heptad repeat regions of gp41, HR1 and HR2, undergo an energetically favorable structural reorganization that results in the formation of a thermostable, six-helix bundle structure that is essential for membrane fusion (Fig. 1D). In the six-helix bundle, three HR2 regions wrap in an antiparallel direction around

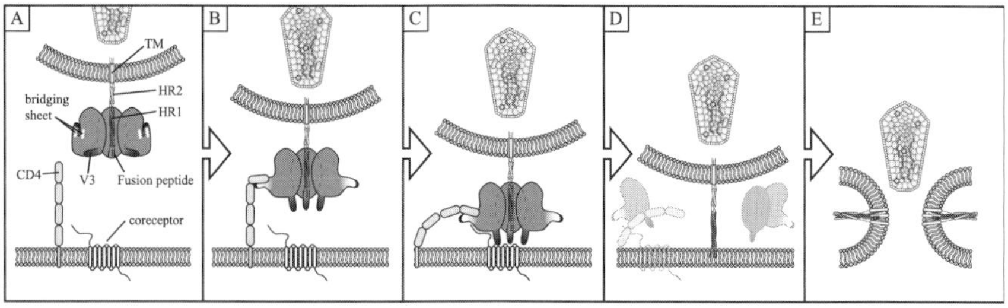

Figure 3. Model of the multi-step entry process that enables HIV to gain access to target cells. (A) The CD4 and coreceptor molecules are embedded in the host membrane (bottom), while the gp120 and gp41 proteins are associated with the viral membrane (curved, top). The V3 loop and bridging sheet domain of gp120 are identified. The gp41 fusion peptide (FP), heptad-repeat (HR1 and HR2), and TM regions are also labeled. (B) The attachment of gp120 to CD4 is associated with conformational changes in gp120 that result in the formation of the bridging sheet domain (white) and the extension of the V3 loop which prior to CD4 binding partially occludes the coreceptor binding site. (C) Coreceptor binding relies on interactions between the bridging sheet and CD4-induced (CD4i) epitopes and the extracellular N' terminal peptide on the coreceptor as well as on interactions between the V3 loop of gp120 and the extracellular loops on the coreceptor. (D) Interactions between gp120, CD4, and coreceptor are believed to result in a conformational change in gp120 that results in dissociation of the envelope trimeric spike, releasing the fusion peptide of gp41, which then inserts into the host membrane. (E) Interaction of the HR1 and HR2 domains of gp41 result in the formation of a six-helix bundle that brings host and viral membranes into close proximity and creates a fusion pore, allowing entry of the HIV capsid into the target cell.

the central coiled-coil of HR1 domains, bringing the N-terminal fusion peptides of the gp41 trimer, which have inserted into the cellular membrane, into close proximity to the TM regions, which traverse the viral membrane [43, 44]. This juxtaposition of the viral and cellular membranes results in the formation of a fusion pore. A schematic model of the multi-step fusion process is presented in Figure 3.

Entry inhibitors

Blocking the interaction between CD4 and gp120 is a logical strategy for preventing HIV infection, although targeting CD4 itself is complicated by side effects due to disruption of CD4 function in immune processes. In contrast, agents that interact with the CD4 binding site on gp120 hold greater promise. One such antiretroviral agent is the small-molecule inhibitor BMS-806, that appears to bind in a pocket in gp120 and either prevents CD4 binding, or prevents CD4-induced conformational changes [45, 46]. However, a major challenge with this class of compounds is the highly variable nature of gp120. It is not uncommon to identify virus strains that are resistant to BMS-806, and those that are sensitive can easily acquire resistance via mutations [47]. More potent, broadly cross-reactive agents are needed if this is to prove to be a viable antiviral strategy. To do this, a structure of unliganded gp120, preferably with the bound drug, may be needed to assist in drug design. A compound with strategic flexibility at specific bonds may be required to enable the inhibitor to adapt to a somewhat variable drug-binding pocket. Until more potent and broadly cross-reactive inhibitors of gp120-CD4 binding are produced, clinical development of this inhibitor class is unlikely to proceed.

Targeting the interaction between virus and the coreceptor molecules is perhaps a more viable strategy for preventing HIV-1 infection of host cells. The CCR5 coreceptor is particularly important for HIV transmission and pathogenesis: the vast majority of virus strains that establish infections in new hosts are those that use CCR5 (R5 strains) [48–53]; the genetic absence of CCR5 results in a high level of protection from HIV infection without significant side effects due to loss of CCR5 function [54–56]; heterozygosity for the inactivating Δ32-ccr5 polymorphism confers a survival advantage upon HIV infection [55–59], indicating that CCR5 levels are rate-limiting for HIV infection *in vivo*; and seven-TM domain receptors are good pharmacological targets. In fact, several CCR5 inhibitors under clinical development and have been shown to reduce viral loads in infected humans [60–62]. Nonetheless, the development of CCR5 inhibitors is not without challenges. Slight variation in the conformation of the helices and extracellular loops of chemokine receptors may result in significant differences in sensitivity to coreceptor inhibitors *in vivo*. Viral resistance to this class of entry inhibitors may occur from either a coreceptor 'switch', either from CCR5 to CXCR4 or *vice versa*, or from altered utilization of the same coreceptor [63–65]. Evidence for both resistance pathways have been found in patients treated with these compounds.

Finally, entry inhibitors targeting the gp41-mediated fusion stage of the entry process have been developed. One of these agents, enfuvirtide, is the only currently approved member of the entry inhibitor class of antiretroviral agents, and is a peptide with an amino acid sequence identical to the HR2 region of gp41. This agent has been demonstrated to potently inhibit HIV infection *in vitro* and *in vivo*, but viral resistance to these compounds has also been identified [66, 67]. Mutations in the HR1 region of gp41 result in decreased sensitivity to enfuvirtide but also result in slower fusion kinetics [68]. *In vivo*, compensatory mutations occur in the HR2 region that improve the kinetics of viral fusion, while maintaining resistance to enfuvirtide [69, 70].

Challenges in the development and use of entry inhibitors

The emerging class of entry inhibitors holds considerable potential for the treatment of patients with HIV infection, particularly those harboring viruses that have resistance to RT and protease inhibitors. However, while progress has been made in understanding the HIV-1 entry process, a number of critical gaps remain. As noted previously, structures of gp120 in an unliganded state and bound to CD4 have been solved, as has the structure of the core of gp41 in the post-fusion state. However, determination of the structures of the gp120-coreceptor interaction, the conformation of gp41 prior to fusion, and the structure of the native trimer remain elusive. Additionally, structures of CD4 inhibitors bound to gp120 and of coreceptor inhibitors bound to CCR5 or CXCR4 are also unavailable. A better understanding of how entry inhibitors bind to Env or coreceptors should make it possible to develop more potent and broadly cross-reactive inhibitors, as well as to design drugs with 'strategic flexibility' that might enable them to bind to a somewhat variable target, such as HIV gp120.

Other challenges in the use of the entry inhibitors are viral and host factors that may alter drug effectiveness *in vivo*. The diversity of the viral envelope proteins suggests that not all viral isolates interact with CD4 and coreceptor in exactly the same way. As a result, there are likely to be some viral isolates that are more sensitive to entry inhibitors and others that are more resistant. Host diversity may also have a role. As indicated by the slower rate of disease progression in patients with the heterozygous Δ32-ccr5 mutation [55, 57–59], the amount of CCR5 expressed on the cell surface is a critical factor in viral pathogenesis. Differences between patients in CCR5 structure or expression levels may also modulate their susceptibility to entry inhibitors [71]. Together, these viral and host factors have a potent effect: viral isolates from patients have differed in susceptibility to enfuvirtide by several orders or magnitude, a much larger range than has been seen with other classes of antiretrovirals [72, 73]. Whether this diversity will affect the clinical outcomes of these patients remains unclear, but must be monitored as use of these agents becomes more established.

Resistance pathways

There are a number of fundamental clinical questions regarding the use of entry inhibitors in the treatment of patients *in vivo*. One of the principal concerns with all antiretroviral agents is the development of viral resistance, and resistance mechanisms to entry inhibitors may not only be complex and variable, but might have the potential to alter viral tropism and pathogenesis by altering the way in which Env binds coreceptors.

In contrast to CD4-binding inhibitors and fusion inhibitors targeting gp41, the coreceptor inhibitors will theoretically be less susceptible to viral resistance mechanisms since they target host proteins rather than the viral envelope. However, the coreceptor inhibitors present some unique challenges also based on the relationship between coreceptor usage, cell tropism, and viral pathogenicity. Coreceptor usage is a principal factor in determining the cellular targets of HIV, with R5-tropic viruses infecting primarily cells of the monocyte and macrophage lineage and memory $CD4^+$ T cells, while X4-tropic viruses predominantly infect naïve $CD4^+$ T cells [74, 75]. X4-tropic viruses are also more infectious and more pathogenic for developing thymocytes than are R5 isolates [76–81]. Viral isolates from patients in early stages of disease are almost universally R5-tropic, regardless of the route of transmission [48–53,

82–85]. Since it appears that a mix of R5 and X4 viruses are transmitted in certain cases [86, 87], and the number of mutations needed to switch coreceptor usage are minimal [88–90], it seems evident that a selection pressure is acting to maintain R5 dominance in early disease. A coreceptor switch from R5- to X4-tropic viruses has been observed in patients during late-stage HIV disease and has been associated with rapid depletion of $CD4^+$ T cells and progression to AIDS [49, 50, 91–95]. However, it remains to be determined whether the emergence of X4-tropic strains is a cause or a consequence of deteriorating immune function.

Viral resistance to coreceptor inhibitors in patients has been seen with two distinct mechanisms. In several patients treated with the R5 inhibitor maraviroc, viral resistance to coreceptor inhibitors has been the result of a 'shift' in viral coreceptor usage from CCR5 to CXCR4 [63]. Notably, the X4-tropic strains that emerged were found to be pre-existing in the patients' viral reservoirs. A second mechanism has been observed in patients treated with other R5 inhibitors, including AD101 and SchD, in which virus continued to utilize the same chemokine receptor but in a drug-insensitive manner [64, 65]. Both of these resistance mechanisms may have profound effects of HIV cell tropism and pathogenicity. Whether a treatment-induced shift from R5- to X4-tropism will accelerate disease progression in patients with preserved immune function is unclear, and will need to be closely monitored during the clinical trials of these agents. The alternative pathway of resistance to coreceptor inhibitors – altered utilization of the same chemokine receptor – may also influence the cellular tropism of HIV. Studies of chemokine receptor mutations that influence sensitivity to AD101 and SchC have suggested that chemokine receptors can exist in several possible conformations on the cell surface [96]. This raises the possibility that altered chemokine receptor usage may influence the subsets of R5- and X4-expressing cells that HIV can infect, potentially changing the pathogenicity of the virus. Future studies of patients developing resistance to coreceptor inhibitors will be important to dissect the mechanisms of viral resistance and their effects on viral pathogenicity and clinical outcome.

Clinical use of HIV entry inhibitors

Although the fusion inhibitor enfuvirtide is the only entry inhibitor currently approved for the treatment of patients, CD4 and coreceptor inhibitors are in various phases of testing (Chapter 10). The varied mechanisms of actions of these agents, acting at different stages of the multi-step entry process, combined with the complications of targeting a highly diverse viral protein with complex resistance pathways, indicates that the effective use of entry inhibitors will require a high degree of clinical acumen. Phenotypic or genotypic tests that predict sensitivity to entry inhibitors – as are available for RT and protease inhibitors – will likely be possible with a better understanding of the mechanisms of viral resistance, and would be useful in selecting agents when initiating therapy or if a change in therapy is required. Similarly, studies will need to be done to address whether combination therapy with several entry inhibitors targeting multiple stages of the entry process may have synergistic effects that may improve viral suppression and reduce side effects. The use of entry inhibitors along with other classes of antiretroviral agents will also have to be investigated. Collectively, the entry inhibitors are a complex but exciting new class of antiretroviral agents that provides significant opportunities and challenges for the treatment of HIV infection and the understanding of HIV pathogenesis.

References

1 Lucas GM, Chaisson RE, Moore RD (1999) Highly active antiretroviral therapy in a large urban clinic: risk factors for virologic failure and adverse drug reactions. *Ann Intern Med* 131: 81–87

2 Yerly S, Kaiser L, Race E, Bru JP, Clavel F, Perrin L (1999) Transmission of antiretroviral-drug-resistant HIV-1 variants. *Lancet* 354: 729–733

3 Chun TW, Davey RT Jr, Engel D, Lane HC, Fauci AS (1999) Re-emergence of HIV after stopping therapy. *Nature* 401: 874–875

4 Finzi D, Hermankova M, Pierson T, Carruth LM, Buck C, Chaisson RE, Quinn TC, Chadwick K, Margolick J, Brookmeyer R et al. (1997) Identification of a reservoir for HIV-1 in patients on highly active antiretroviral therapy. *Science* 278: 1295–1300

5 Starcich BR, Hahn BH, Shaw GM, McNeely PD, Modrow S, Wolf H, Parks ES, Parks WP, Josephs SF, Gallo RC, Wong-Staal F (1986) Identification and characterization of conserved and variable regions in the envelope gene of HTLV-III/LAV, the retrovirus of AIDS. *Cell* 45: 637–648

6 Leonard CK, Spellman MW, Riddle L, Harris RJ, Thomas JN, Gregory TJ (1990) Assignment of intrachain disulfide bonds and characterization of potential glycosylation sites of the type 1 recombinant human immunodeficiency virus envelope glycoprotein (gp120) expressed in Chinese hamster ovary cells. *J Biol Chem* 265: 10373–10382

7 Kwong PD, Wyatt R, Robinson J, Sweet RW, Sodroski J, Hendrickson WA (1998) Structure of an HIV gp120 envelope glycoprotein in complex with the CD4 receptor and a neutralizing human antibody. *Nature* 393: 648–659

8 Huang CC, Tang M, Zhang MY, Majeed S, Montabana E, Stanfield RL, Dimitrov DS, Korber B, Sodroski J, Wilson IA et al. (2005) Structure of a V3-containing HIV-1 gp120 core. *Science* 310: 1025–1028

9 Chen B, Vogan EM, Gong H, Skehel JJ, Wiley DC, Harrison SC (2005) Structure of an unliganded simian immunodeficiency virus gp120 core. *Nature* 433: 834–841

10 Rizzuto CD, Wyatt R, Hernandez-Ramos N, Sun Y, Kwong PD, Hendrickson WA, Sodroski J (1998) A conserved HIV gp120 glycoprotein structure involved in chemokine receptor binding. *Science* 280: 1949–1953

11 Dubay JW, Roberts SJ, Brody B, Hunter E (1992) Mutations in the leucine zipper of the human immunodeficiency virus type 1 transmembrane glycoprotein affect fusion and infectivity. *J Virol* 66: 4748–4756

12 Wild C, Dubay JW, Greenwell T, Baird T Jr, Oas TG, McDanal C, Hunter E, Matthews T (1994) Propensity for a leucine zipper-like domain of human immunodeficiency virus type 1 gp41 to form oligomers correlates with a role in virus- induced fusion rather than assembly of the glycoprotein complex. *Proc Natl Acad Sci USA* 91: 12676–12680

13 Wild CT, Shugars DC, Greenwell TK, McDanal CB, Matthews TJ (1994) Peptides corresponding to a predictive alpha-helical domain of human immunodeficiency virus type 1 gp41 are potent inhibitors of virus infection. *Proc Natl Acad Sci USA* 91: 9770–9774

14 Zhu P, Chertova E, Bess J Jr, Lifson JD, Arthur LO, Liu J, Taylor KA, Roux KH (2003) Electron tomography analysis of envelope glycoprotein trimers on HIV and simian immunodeficiency virus virions. *Proc Natl Acad Sci USA* 100: 15812–15817

15 Zhu P, Liu J, Bess J Jr, Chertova E, Lifson JD, Grise H, Ofek GA, Taylor KA, Roux KH (2006)

Distribution and three-dimensional structure of AIDS virus envelope spikes. *Nature* 441: 847–852
16 Trkola A, Dragic T, Arthos J, Binlay JM, Olson WC, Allaway GP, Cheng-Meyer C, Robinson J, Maddon PJ, Moore JP (1996) CD4-dependent, antibody-sensitive interactions between HIV-1 and its co-receptor CCR-5. *Nature* 384: 184–187
17 Wu L, Gerard NP, Wyatt R, Choe H, Parolin C, Ruffing N, Borsetti A, Cardoso AA, Desjardin E, Newman W, Gerard C, Sodroski J (1996) CD4-induced interaction of primary HIV-1 gp120 glycoproteins with the chemokine receptor CCR-5. *Nature* 384: 179–183
18 Zhang YJ, Dragic T, Cao Y, Kostrikis L, Kwon DS, Littman DR, KewalRamani VN, Moore JP (1998) Use of coreceptors other than CCR5 by non-syncytium-inducing adult and pediatric isolates of human immunodeficiency virus type 1 is rare *in vitro*. *J Virol* 72: 9337–9344
19 Dragic T, Litwin V, Allaway GP, Martin SR, Huang Y, Nagashima KA, Cayanan C, Maddon PJ, Koup RA, Moore JP, Paxton WA (1996) HIV-1 entry into $CD4^+$ cells is mediated by the chemokine receptor CC-CKR-5. *Nature* 381: 667–673
20 Doranz BJ, Rucker J, Yi Y, Smyth RJ, Samson M, Peiper SC, Parmentier M, Collman RG, Doms RW (1996) A dual-tropic primary HIV-1 isolate that uses fusin and the beta- chemokine receptors CKR-5, CKR-3, and CKR-2b as fusion cofactors. *Cell* 85: 1149–1158
21 Choe H, Farzan M, Sun Y, Sullivan N, Rollins B, Ponath PD, Wu L, Mackay CR, LaRosa G, Newman W et al. (1996) The beta-chemokine receptors CCR3 and CCR5 facilitate infection by primary HIV-1 isolates. *Cell* 85: 1135–1148
22 Alkhatib G, Combadiere C, Broder CC, Feng Y, Kennedy PE, Murphy PM, Berger EA (1996) CC CKR5: a RANTES, MIP-1alpha, MIP-1beta receptor as a fusion cofactor for macrophage-tropic HIV-1. *Science* 272: 1955–1958
23 Deng H, Liu R, Ellmeier W, Choe S, Unutmaz D, Burkhart M, Di Marzio P, Marmon S, Sutton RE, Hill CM et al. (1996) Identification of a major co-receptor for primary isolates of HIV-1. *Nature* 381: 661–666
24 Oberlin E, Amara A, Bachelerie F, Bessia C, Virelizier JL, Arenzana-Seisdedos F, Schwartz O, Heard JM, Clark-Lewis I, Legler DF et al. (1996) The CXC chemokine SDF-1 is the ligand for LESTR/fusin and prevents infection by T-cell-line-adapted HIV-1. *Nature* 382: 833–835
25 Feng Y, Broder CC, Kennedy PE, Berger EA (1996) HIV-1 entry cofactor: functional cDNA cloning of a seven-transmembrane, G protein-coupled receptor. *Science* 272: 872–877
26 Basmaciogullari S, Babcock GJ, Van Ryk D, Wojtowicz W, Sodroski J (2002) Identification of conserved and variable structures in the human immunodeficiency virus gp120 glycoprotein of importance for CXCR4 binding. *J Virol* 76: 10791–10800
27 Hoffman TL, LaBranche CC, Zhang W, Canziani G, Robinson J, Chaiken I, Hoxie JA, Doms RW (1999) Stable exposure of the coreceptor-binding site in a CD4-independent HIV- 1 envelope protein. *Proc Natl Acad Sci USA* 96: 6359–6364
28 Cormier EG, Tran DN, Yukhayeva L, Olson WC, Dragic T (2001) Mapping the determinants of the CCR5 amino-terminal sulfopeptide interaction with soluble human immunodeficiency virus type 1 gp120-CD4 complexes. *J Virol* 75: 5541–5549
29 Cormier EG, Dragic T (2002) The crown and stem of the V3 loop play distinct roles in human immunodeficiency virus type 1 envelope glycoprotein interactions with the CCR5 coreceptor. *J Virol* 76: 8953–8957
30 Rizzuto C, Sodroski J (2000) Fine definition of a conserved CCR5-binding region on the human immunodeficiency virus type 1 glycoprotein 120. *AIDS Res Hum Retroviruses* 16: 741–749
31 Wu L, LaRosa G, Kassam N, Gordon CJ, Heath H, Ruffing N, Chen H, Humblias J, Samson M, Parmentier M, Moore JP, Mackay CR (1997) Interaction of chemokine receptor CCR5 with its ligands: multiple domains for HIV-1 gp120 binding and a single domain for chemokine binding. *J Exp Med* 186: 1373–1381
32 Lee B, Sharron M, Blanpain C, Doranz BJ, Vakili J, Setoh P, Berg E, Liu G, Guy HR, Durell SR et al. (1999) Epitope mapping of CCR5 reveals multiple conformational states and distinct but overlapping structures involved in chemokine and coreceptor function. *J Biol Chem* 274: 9617–9626
33 Samson M, LaRosa G, Libert F, Paindavoine P, Detheux M, Vassart G, Parmentier M (1997) The second extracellular loop of CCR5 is the major determinant of ligand specificity. *J Biol Chem* 272: 24934–24941
34 Platt EJ, Kuhmann SE, Rose PP, Kabat D (2001) Adaptive mutations in the V3 loop of gp120 enhance fusogenicity of human immunodeficiency virus type 1 and enable use of a CCR5 coreceptor that lacks the amino-terminal sulfated region. *J Virol* 75: 12266–12278

35 Farzan M, Mirzabekov T, Kolchinksy P, Wyatt R, Cabyabyab M, Gerard NP, Gerard C, Sodroski J, Choe H (1999) Tyrosine sulfation of the amino terminus of CCR5 facilitates HIV-1 entry. *Cell* 96: 667–676
36 Farzan M, Vasilieva N, Schnitzler CE, Chung S, Robinson J, Gerard NP, Gerard C, Choe H, Sodroski J (2000) A tyrosine-sulfated peptide based on the N terminus of CCR5 interacts with a CD4-enhanced epitope of the HIV-1 gp120 envelope glycoprotein and inhibits HIV-1 entry. *J Biol Chem* 275: 33516–33521
37 Cormier EG, Persuh M, Thompson DA, Lin SW, Sakmar TP, Olson WC, Dragic T (2000) Specific interaction of CCR5 amino-terminal domain peptides containing sulfotyrosines with HIV-1 envelope glycoprotein gp120. *Proc Natl Acad Sci USA* 97: 5762–5767
38 Lin G, Baribaud F, Romano J, Doms RW, Hoxie JA (2003) Identification of gp120 binding sites on CXCR4 by using CD4-independent human immunodeficiency virus type 2 Env proteins. *J Virol* 77: 931–942
39 Lu Z, Berson JF, Chen Y, Turner JD, Zhang T, Sharron M, Jenks MH, Wang Z, Kim J, Rucker J, Hoxie JA, Peiper SC, Doms RW (1997) Evolution of HIV-1 coreceptor usage through interactions with distinct CCR5 and CXCR4 domains. *Proc Natl Acad Sci USA* 94: 6426–6431
40 Chabot DJ, Zhang PF, Quinnan GV, Broder CC (1999) Mutagenesis of CXCR4 identifies important domains for human immunodeficiency virus type 1 X4 isolate envelope-mediated membrane fusion and virus entry and reveals cryptic coreceptor activity for R5 isolates. *J Virol* 73: 6598–6609
41 Doranz BJ, Orsini MJ, Turner JD, Hoffman TL, Berson JF, Hoxie JA, Peiper SC, Brass LF, Doms RW (1999) Identification of CXCR4 domains that support coreceptor and chemokine receptor functions. *J Virol* 73: 2752–2761
42 Abrahamyan LG, Markosyan RM, Moore JP, Cohen FS, Melikyan GB (2003) Human immunodeficiency virus type 1 Env with an intersubunit disulfide bond engages coreceptors but requires bond reduction after engagement to induce fusion. *J Virol* 77: 5829–5836
43 Chan DC, Fass D, Berger JM, Kim PS (1997) Core structure of gp41 from the HIV envelope glycoprotein. *Cell* 89: 263–273
44 Weissenhorn W, Dessen A, Harrison SC, Skehel JJ, Wiley DC (1997) Atomic structure of the ectodomain from HIV-1 gp41. *Nature* 387: 426–430
45 Lin PF, Blair W, Wang T, Spicer T, Guo Q, Zhou N, Gong YF, Wang HG, Rose R, Yamanaka G et al. (2003) A small molecule HIV-1 inhibitor that targets the HIV-1 envelope and inhibits CD4 receptor binding. *Proc Natl Acad Sci USA* 100: 11013–11018
46 Si Z, Madani N, Cox JM, Chruma JJ, Klein JC, Schon A, Phan N, Wang L, Biorn AC, Cocklin S et al. (2004) Small-molecule inhibitors of HIV-1 entry block receptor-induced conformational changes in the viral envelope glycoproteins. *Proc Natl Acad Sci USA* 101: 5036–5041
47 Madani N, Perdigoto AL, Srinivasan K, Cox JM, Chruma JJ, LaLonde J, Head M, Smith AB 3rd, Sodroski JG (2004) Localized changes in the gp120 envelope glycoprotein confer resistance to human immunodeficiency virus entry inhibitors BMS-806 and #155. *J Virol* 78: 3742–3752
48 Balotta C, Vigano A, Riva C, Colombo MC, Salvaggio A, de Pasquale MP, Crupi L, Papagno L, Galli M, Moroni M, Principi N (1996) HIV type 1 phenotype correlates with the stage of infection in vertically infected children. *AIDS Res Hum Retroviruses* 12: 1247–1253
49 Schuitemaker H, Koot M, Kootstra NA, Dercksen MW, de Goede RE, van Steenwijk RP, Lange JM, Schattenkerk JK, Miedema F, Tersmette M (1992) Biological phenotype of human immunodeficiency virus type 1 clones at different stages of infection: progression of disease is associated with a shift from monocytotropic to T-cell-tropic virus population. *J Virol* 66: 1354–1360
50 Connor RI, Sheridan KE, Ceradini D, Choe S, Landau NR (1997) Change in coreceptor use coreceptor use correlates with disease progression in HIV-1 infected individuals. *J Exp Med* 185: 621–628
51 Salvatori F, Scarlatti G (2001) HIV type 1 chemokine receptor usage in mother-to-child transmission. *AIDS Res Hum Retroviruses* 17: 925–935
52 Long EM, Rainwater SM, Lavreys L, Mandaliya K, Overbaugh J (2002) HIV type 1 variants transmitted to women in Kenya require the CCR5 coreceptor for entry, regardless of the genetic complexity of the infecting virus. *AIDS Res Hum Retroviruses* 18: 567–576
53 Casper CH, Clevestig P, Carlenor E, Leitner T, Anzen B, Lidman K, Belfrage E, Albert J, Bohlin AB, Naver L et al. (2002) Link between the X4 phenotype in human immunodeficiency virus type 1-infected mothers and their children, despite the early presence of R5 in the child. *J Infect Dis* 186: 914–921

54 Liu R, Paxton WA, Choe S, Ceradini D, Martin SR, Horuk R, MacDonald ME, Stuhlmann H, Koup RA, Landau NR (1996) Homozygous defect in HIV-1 coreceptor accounts for resistance of some multiply-exposed individuals to HIV-1 infection. *Cell* 86: 367–377

55 Dean M, Carrington M, Winkler C, Huttley GA, Smith MW, Allikmets R, Goedert JJ, Buchbinder SP, Vittinghoff E, Gomperts E et al. (1996) Genetic restriction of HIV-1 infection and progression to AIDS by a deletion allele of the CKR5 structural gene. Hemophilia Growth and Development Study, Multicenter AIDS Cohort Study, Multicenter Hemophilia Cohort Study, San Francisco City Cohort, ALIVE Study. *Science* 273: 1856–1862

56 Samson M, Libert F, Doranz BJ, Rucker J, Liesnard C, Farber CM, Saragosti S, Lapoumeroulie C, Cognaux J, Forceille C et al. (1996) Resistance to HIV-1 infection in Caucasian individuals bearing mutant alleles of the CCR-5 chemokine receptor gene. *Nature* 382: 722–725

57 Huang Y, Paxton WA, Wolinsky SM, Neumann AU, Zhang L, He T, Kang S, Ceradini D, Jin Z, Yazdanbakhsh K et al. (1996) The role of a mutant CCR5 allele in HIV-1 transmission and disease progression. *Nat Med* 2: 1240–1243

58 Rappaport J, Cho YY, Hendel H, Schwartz EJ, Schachter F, Zagury JF (1997) 32 bp CCR-5 gene deletion and resistance to fast progression in HIV-1 infected heterozygotes. *Lancet* 349: 922–923

59 Michael NL, Chang G, Louie LG, Mascola JR, Dondero D, Birx DL, Sheppard HW (1997) The role of viral phenotype and CCR-5 gene defects in HIV-1 transmission and disease progression. *Nat Med* 3: 338–340

60 Reynes J, Rouzie R, Kanouni T, Baillat V, Baroudy B, Keung A, Hogan C, Markowitz M, Laughlin M (2002) SCH C: Safety and antiviral effects of a CCR5 receptor antagonist in HIV-1 infected subjects. In: *9th Conference on Retroviruses and Opportunistic Infections*, Seattle, WA. Abstract no. 1

61 Pozniak AL, Fätkenheuer G, Johnson M, Hoepelman IM, Rockstroh J, Goebel F, Abel S, James I, Rosario M, Medhurst C et al. (2003) Effect of short-term monotherapy with UK-427,857 on viral load in HIV-infected patients. In: *43rd Interscience conference on Antimicrobial Agents and Chemotherapy*, Chicago, IL. Abstract no. H-443

62 Schurmann D, Rouzier R, Nougarede R, Reynes J, Fatkenheuer G, Raffi F, Michelet C, Tarral A, Hoffmann C, Kiunke J, Sprenger H, vanLier J, Sansone A, Jackson M, Lauglin M (2004) Antiviral activity of a CCR5 receptor antagonist. In: *11th Conference on Retroviruses and Opportunistic Infections*. Abstract no. 140LB

63 Westby M, Lewis M, Whitcomb J, Youle M, Pozniak AL, James IT, Jenkins TM, Perros M., van der Ryst E (2006) Emergence of CXCR4-using human immunodeficiency virus type 1 (HIV-1) variants in a minority of HIV-1-infected patients following treatment with the CCR5 antagonist maraviroc is from a pretreatment CXCR4-using virus reservoir. *J Virol* 80: 4909–4920

64 Marozsan AJ, Kuhmann SE, Morgan T, Herrera C, Rivera-Troche E, Xu S, Baroudy BM, Strizki J, Moore JP (2005) Generation and properties of a human immunodeficiency virus type 1 isolate resistant to the small molecule CCR5 inhibitor, SCH-417690 (SCH-D). *Virology* 338: 182–199

65 Trkola A, Kuhmann SE, Strizki JM, Maxwell E, Ketas T, Morgan T, Pugach P, Xu S, Wojcik L, Tagat J et al. (2002) HIV-1 escape from a small molecule, CCR5-specific entry inhibitor does not involve CXCR4 use. *Proc Natl Acad Sci USA* 99: 395–400

66 Kilby JM, Hopkins S, Venetta TM, DiMassimo B, Cloud GA, Lee JY, Alldredge L, Hunter E, Lambert D, Bolognesi D et al. (1998) Potent suppression of HIV-1 replication in humans by T-20, a peptide inhibitor of gp41-mediated virus entry. *Nat Med* 4: 1302–1307

67 Wei X, Decker JM, Liu H, Zhang Z, Arani RB, Kilby JM, Saag MS, Wu X, Shaw GM, Kappes JC (2002) Emergence of resistant human immunodeficiency virus type 1 in patients receiving fusion inhibitor (T-20) monotherapy. *Antimicrob Agents Chemother* 46: 1896–1905

68 Reeves JD, Lee FH, Miamidian JL, Jabara CB, Juntilla MM, Doms RW (2005) Enfuvirtide resistance mutations: impact on human immunodeficiency virus envelope function, entry inhibitor sensitivity, and virus neutralization. *J Virol* 79: 4991–4999

69 Menzo S, Castagna A, Monachetti A, Hasson H, Danise A, Carini E, Bagnarelli P, Lazzarin A, Clementi M (2004) Resistance and replicative capacity of HIV-1 strains selected *in vivo* by long-term enfuvirtide treatment. *New Microbiol* 27: 51–61

70 Neumann T, Hagmann I, Lohrengel S, Heil ML, Derdeyn CA, Krausslich HG, Dittmar MT (2005) T20-insensitive HIV-1 from naive patients exhibits high viral fitness in a novel dual-color competition assay on primary cells. *Virology* 333: 251–262

71 Reeves JD, Gallo SA, Ahmad N, Miamidian JL, Harvey PE, Sharron M, Pohlmann S, Sfakianos JN, Derdeyn CA, Blumenthal R et al. (2002) Sensitivity of HIV-1 to entry inhibitors correlates

with envelope/coreceptor affinity, receptor density, and fusion kinetics. *Proc Natl Acad Sci USA* 99: 16249–16254

72 Derdeyn CA, Decker JM, Sfakianos JN, Wu X, O'Brien WA, Ratner L, Kappes JC, Shaw GM, Hunter E (2000) Sensitivity of human immunodeficiency virus type 1 to the fusion inhibitor T-20 is modulated by coreceptor specificity defined by the V3 loop of gp120. *J Virol* 74: 8358–8367

73 Melby T, Despirito M, Demasi R, Heilek-Snyder G, Greenberg ML, Graham N (2006) HIV-1 coreceptor use in triple-class treatment-experienced patients: Baseline prevalence, correlates, and relationship to enfuvirtide response. *J Infect Dis* 194: 238–246

74 Bleul CC, Wu L, Hoxie JA, Springer TA, Mackay CR (1997) The HIV coreceptors CXCR4 and CCR5 are differentially expressed and regulated on human T lymphocytes. *Proc Natl Acad Sci USA* 94: 1925–1930

75 Blaak H, van't Wout AB, Brouwer M, Hooibrink B, Hovenkamp E, Schuitemaker H (2000) *In vivo* HIV-1 infection of CD45RA(+)CD4(+) T cells is established primarily by syncytium-inducing variants and correlates with the rate of CD4(+) T cell decline. *Proc Natl Acad Sci USA* 97: 1269–1274

76 Jamieson BD, Pang S, Aldrovandi GM, Zha J, Zack JA (1995) *In vivo* pathogenic properties of two clonal human immunodeficiency virus type 1 isolates. *J Virol* 69: 6259–6264

77 Berkowitz RD, Alexander S, McCune JM (2000) Causal relationships between HIV-1 coreceptor utilization, tropism, and pathogenesis in human thymus. *AIDS Res Hum Retroviruses* 16: 1039–1045

78 Berkowitz RD, Alexander S, Bare C, Linquist-Stepps V, Bogan M, Moreno ME, Gibson L, Wieder ED, Kosek J, Stoddart CA, McCune JM (1998) CCR5- and CXCR4-utilizing strains of human immunodeficiency virus type 1 exhibit differential tropism and pathogenesis *in vivo*. *J Virol* 72: 10108–10117

79 Berkowitz RD, van't Wout AB, Kootstra NA, Moreno ME, Linquist-Stepps VD, Bare C, Stoddart CA, Schuitemaker H, McCune JM (1999) R5 strains of human immunodeficiency virus type 1 from rapid progressors lacking X4 strains do not possess X4-type pathogenicity in human thymus. *J Virol* 73: 7817–7822

80 Correa R, Munoz-Fernandez MA (2001) Viral phenotype affects the thymic production of new T cells in HIV-1-infected children. *AIDS* 15: 1959–1963

81 Kaneshima H, Su L, Bonyhadi ML, Connor RI, Ho DD, McCune JM (1994) Rapid-high, syncytium-inducing isolates of human immunodeficiency virus type 1 induce cytopathicity in the human thymus of the SCID-hu mouse. *J Virol* 68: 8188–8192

82 Roos MT, Lange JM, de Goede RE, Coutinho RA, Schellekens PT, Miedema F, Tersmette M (1992) Viral phenotype and immune response in primary human immunodeficiency virus type 1 infection. *J Infect Dis* 165: 427–432

83 van't Wout AB, Kootstra NA, Mulder-Kampinga GA, Albrecht-van Lent N, Scherpbier HJ, Veenstra J, Boer K, Coutinho RA, Miedema F, Schuitemaker H (1994) Macrophage-tropic variants initiate human immunodeficiency virus type 1 infection after sexual, parenteral, and vertical transmission. *J Clin Invest* 94: 2060–2067

84 Zhu T, Mo H, Wang N, Nam DS, Cao Y, Koup RA, Ho DD (1993) Genotypic and phenotypic characterization of HIV-1 patients with primary infection. *Science* 261: 1179–1181

85 Zhang LQ, MacKenzie P, Cleland A, Holmes EC, Brown AJ, Simmonds P (1993) Selection for specific sequences in the external envelope protein of human immunodeficiency virus type 1 upon primary infection. *J Virol* 67: 3345–3356

86 Cornelissen M, Mulder-Kampinga G, Veenstra J, Zorgdrager F, Kuiken C, Hartman S, Dekker J, van der Hoek L, Sol C, Coutinho R (1995) Syncytium-inducing (SI) phenotype suppression at seroconversion after intramuscular inoculation of a non-syncytium-inducing/SI phenotypically mixed human immunodeficiency virus population. *J Virol* 69: 1810–1818

87 Pratt RD, Shapiro JF, McKinney N, Kwok S, Spector SA (1995) Virologic characterization of primary human immunodeficiency virus type 1 infection in a health care worker following needlestick injury. *J Infect Dis* 172: 851–854

88 Fouchier RA, Groenink M, Kootstra NA, Tersmette M, Huisman HG, Miedema F, Schuitemaker H (1992) Phenotype-associated sequence variation in the third variable domain of the human immunodeficiency virus type 1 gp120 molecule. *J Virol* 66: 3183–3187

89 De Jong JJ, De Ronde A, Keulen W, Tersmette M, Goudsmit J (1992) Minimal requirements for the human immunodeficiency virus type 1 V3 domain to support the syncytium-inducing phenotype: analysis by single amino acid substitution. *J Virol* 66: 6777–6780

90 Shioda T, Levy JA, Cheng-Mayer C (1992) Small amino acid changes in the V3 hypervariable region of gp120 can affect the T-cell-line and macrophage tropism of human immunodeficiency virus type 1. *Proc Natl Acad Sci USA* 89: 9434–9438
91 de Roda Husman AM, Koot M, Cornelissen M, Keet IP, Brouwer M, Broersen SM, Bakker M, Roos MT, Prins M, de Wolf F et al. (1997) Association between CCR5 genotype and the clinical course of HIV-1 infection. *Ann Intern Med* 127: 882–890
92 Karlsson A, Parsmyr K, Sandstrom E, Fenyo EM, Albert J (1994) MT-2 cell tropism as prognostic marker for disease progression in human immunodeficiency virus type 1 infection. *J Clin Microbiol* 32: 364–370
93 Maas JJ, Gange SJ, Schuitemaker H, Coutinho RA, van Leeuwen R, Margolick JB (2000) Strong association between failure of T cell homeostasis and the syncytium-inducing phenotype among HIV-1-infected men in the Amsterdam Cohort Study. *AIDS* 14: 1155–1161
94 Richman DD, Bozzette SA (1994) The impact of the syncytium-inducing phenotype of human immunodeficiency virus on disease progression. *J Infect Dis* 169: 968–974
95 Scarlatti G, Tresoldi E, Bjorndal A, Fredriksson R, Colognesi C, Deng HK, Malnati MS, Plebani A, Siccardi AG, Littman DR et al. (1997) *In vivo* evolution of HIV-1 co-receptor usage and sensitivity to chemokine-mediated suppression. *Nat Med* 3: 1259–1265
96 Billick E, Seibert C, Pugach P, Ketas T, Trkola A, Endres MJ, Murgolo NJ, Coates E, Reyes GR, Baroudy BM et al. (2004) The differential sensitivity of human and rhesus macaque CCR5 to small-molecule inhibitors of human immunodeficiency virus type 1 entry is explained by a single amino acid difference and suggests a mechanism of action for these inhibitors. *J Virol* 78: 4134–4144

Entry Inhibitors in HIV Therapy
Edited by Jacqueline D. Reeves and Cynthia A. Derdeyn

The challenge of HIV sequence diversity in the envelope glycoprotein

Laurence Vergne † and Martine Peeters

UMR145 "HIV and associated diseases", Institut de Recherche pour le Dévelopement, 911 av Agropolis, 34394 Montpellier, France

One of the major characteristics of the human immunodeficiency viruses (HIVs) is their extremely high genetic variability, which makes HIV one of the fastest evolving among many other human pathogens. Viral heterogeneity is one of the classical means by which HIV evades the host immune system, and also leads to the resistance to various antiretroviral regimens. Highly active antiretroviral therapy to treat HIV-infected patients is mainly based on a combination of protease and reverse transcriptase inhibitors. However, resistance to these drugs, the persistence of latent viral reservoirs and long-term toxicities require the continuous need to improve these classes of drugs and to develop new drugs with other targets. Therefore, a new generation of drugs has been recently developed to inhibit viral entry into the cell.

This chapter presents obstacles and considerations for HIV entry inhibition that can result from the high levels of sequence diversity observed in Env glycoproteins compared to other antiretroviral targets, like protease and RT sequences, which have relatively well-conserved sequences.

Why does HIV have high genetic diversity?

The extensive heterogeneity of HIV is the result of several factors. First, there have been multiple introductions of the genetically diverse simian viruses into humans. Both HIV-1 and HIV-2 represent zoonotic transmissions from two different sources, namely chimpanzees (*Pan troglodytes*) and sooty mangabeys (*Cercocebus atys*). Second, since these simian viruses entered the human population, they have rapidly accumulated more genetic diversity because of the high error rate and recombinogenic properties of reverse transcriptase (RT), and the fast turnover of virions in HIV-infected individuals.

The RT enzyme, which is responsible for converting the single-stranded viral genomic RNA to double-stranded DNA that integrates into the host chromosome, is known for its high rate of incorrect nucleotide substitutions during this reverse transcription step: at least 0.2 errors per genome during each replication cycle [1, 2]. As reverse transcription is the first step of the viral replication cycle, errors made at this stage ensure propagation of the erroneously copied genome to form the quasi-species of HIV found in the infected individuals. Among these quasi-species, between 1% and 5% diversity can be observed in some viral genes, and each quasi-species continues to propagate the cycle of error-prone reverse transcription and recombination. Each HIV-1-infected patient is thus infected with a population of highly related but genetically distinct viruses. Figure 1 shows the intra-patient diversity for different subtype- (three subtypes C and three subtypes E) infected patients at two different time points. For each patient, a high level of diversity is seen at each time point, and moreover, over time (13–22 months later) diversity increases, and many of the previously documented variants are no longer present.

This rapid variability provides the virus with the capability to adapt its genome to escape selective pressures from the immune system and antiretroviral therapy. The high genetic variability of HIV is also the result of the recombinogenic properties of RT [3]. HIV, like all other retroviruses, is diploid, containing two genomic RNA molecules per virion. Therefore, cells infected with two different HIV strains [subtypes, circulating recombinant forms (CRFs) or quasi-species] might produce heterozygous virions, providing an opportunity for recombination to occur during reverse transcription. HIV-1 recombines approximately two to three times per genome per replication cycle [4].

The rapid viral turnover in an HIV-infected individual also plays a role in the extent of heterogeneity [5]. It has been estimated that about 300 rounds of replication occur per year in infected patients [6]; each day about 10^{10} to 10^{12} new virions are produced [7].

Classification of HIV and geographic distribution of HIV-1 variants

Classification of HIV

On the basis of phylogenetic analyses of numerous isolates obtained from diverse geographic origins, HIV is subdivided into types (HIV-1 and HIV-2), groups, subtypes, sub-subtypes, CRFs and unique recombinants (URFs) [8]. The initial epicenters of HIV-1 and HIV-2 infection appear to have been Central Africa and West Africa, respectively, reflecting the natural habitats of chimpanzees and sooty mangabeys. HIV-2 is still primarily found in West Africa, and currently the HIV-2 prevalence is stable or even decreasing [9]. In contrast, HIV-1 has spread throughout Africa, including West Africa, and some lineages of HIV-1 have dispersed around the world, so that HIV-1 is predominant globally.

Three HIV-1 groups (M, N, O) exist. Group M (for major) represents the vast majority of HIV-1 strains found worldwide and is responsible for the pandemic [10]. Variability between the three groups is estimated at 30% overall,

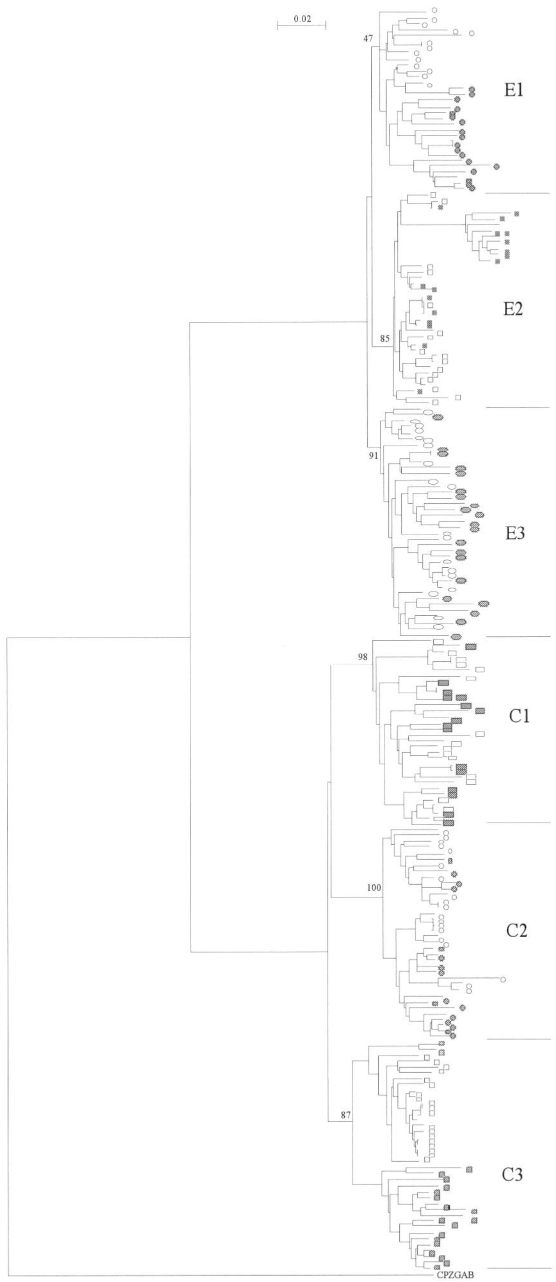

Figure 1. HIV gp120 intra-patient variability for six patients infected with different HIV-1 subtypes and characterized at different time points. Three patients were infected with subtype C (C1, C2, and C3) and three by subtype E (E1, E2, and E3) (called CRF01_AE). Genetic characterization (V1–V5) was performed at twice (black and gray), with an period of 13–22 months between samples, for each patient.

and 50% for the *env* gene. HIV-1 groups O and N are genetically very divergent from group M. Group O is endemic to Cameroon and neighboring countries in West Central Africa, but even there these viruses represent a minority of HIV-1 strains: their highest prevalence is 1–5% of HIV-1-positive samples [11]. Phylogenetic analyses of group O strains have not revealed the same substructure as that found within the evolutionary tree of group M. At present, it is unclear whether group O should be divided into subtypes because only a limited number of full genomes are available, which do not describe the full spectrum of group O diversity that is suggested through analysis of partial genome sequences [12]. Group N forms an independent lineage most closely related to group M, using the sequence from the 5' end of the genome, and clusters more closely with a chimpanzee virus (SIVcpz), using the sequence from the 3' end of the genome [13]. These viruses have only been identified in Cameroon, and represent only a minority of HIV-1 infections, with about 10 patients identified to date.

Group M can be further subdivided in subtypes (A–D, F–H, J and K), subsubtypes (A1 and A4; F1 and F2), and mosaic viruses, called CRFs when they play a major role in an epidemic (Fig. 2). Currently, more than 30 different CRFs have been reported. Figure 3 shows the complex recombinant structures

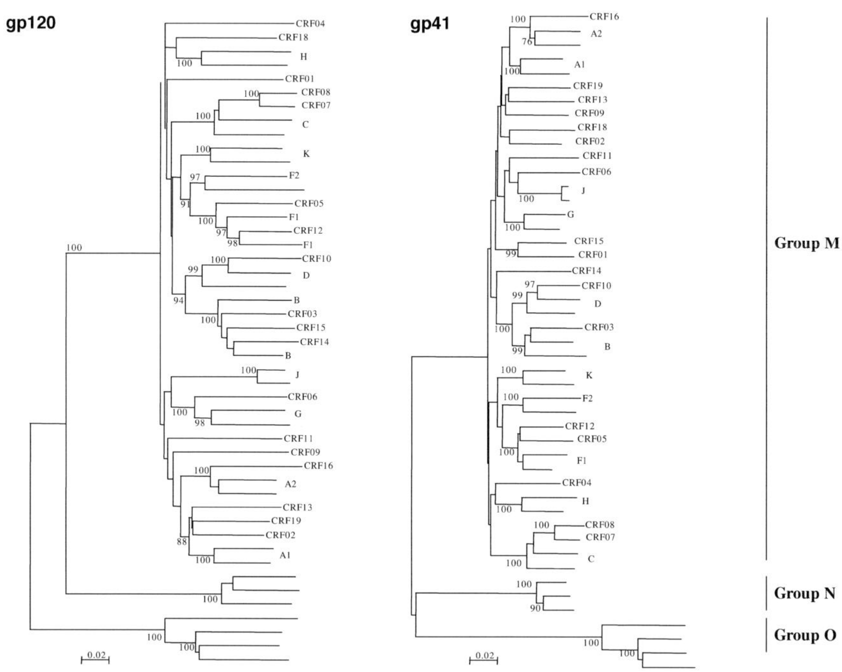

Figure 2. Phylogenetic tree analysis of gp120 and gp41.

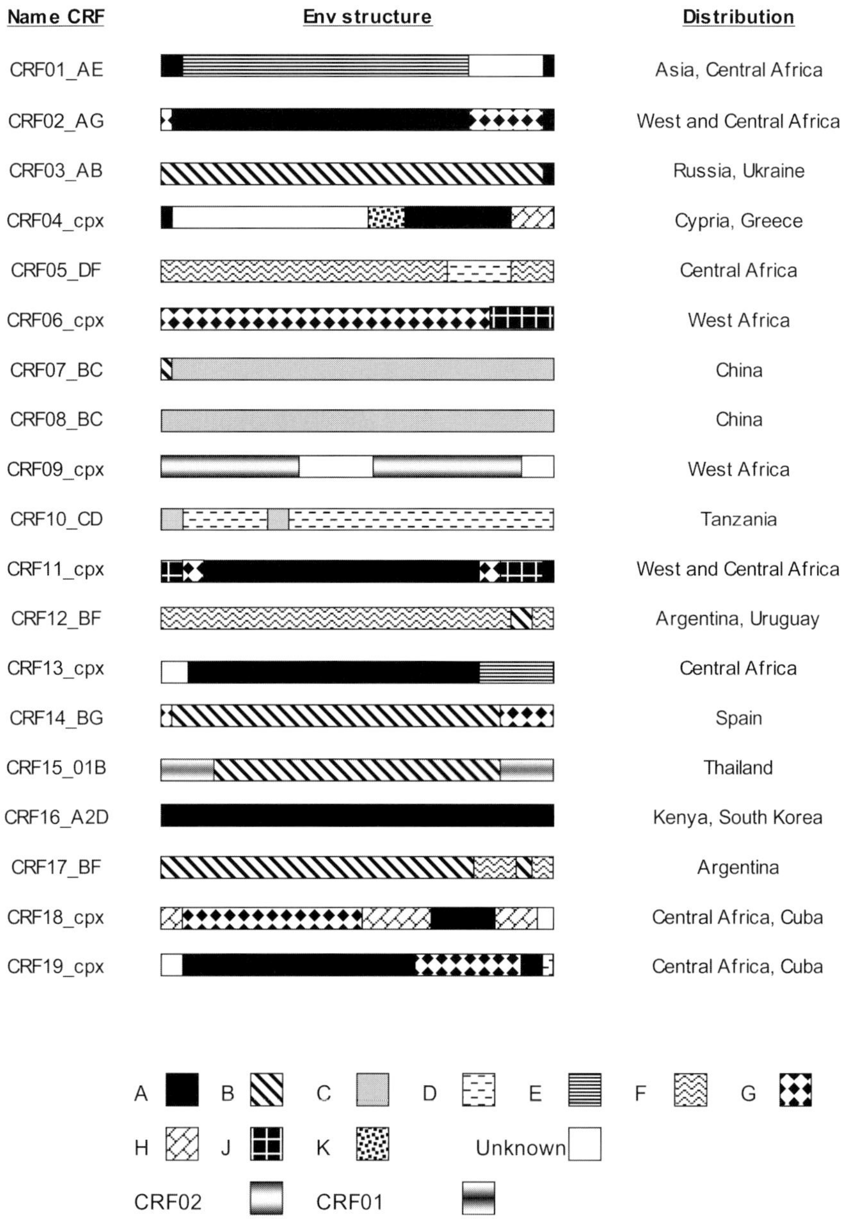

Figure 3. Schematic representation of the env mosaic structure of circulating recombinant forms (CRFs).

for 19 CRFs in the *env* gene. The variability between subtypes within M varies for different genes; inter-subtype nucleotide sequence divergence may exceed 20%, 15% and 30% for *gag*, *pol*, and *env*, respectively. The subtypes are approximately equidistant from each other over their entire genome and, by definition, CRFs should resemble each other also over the entire genome, with similar breakpoints reflecting common ancestry from the same recombination event. In addition, many unique recombinants have been documented.

Worldwide distribution of HIV-1 group M variants

Subtype and CRF designations have been powerful molecular epidemiological markers to track the course of the HIV-1 pandemic. Extensive efforts have been made to collect and characterize HIV-1 isolates from around the world, and a broad picture of the distribution of HIV-1 variants has emerged (Fig. 4). Globally, the predominant viral forms are subtypes A and C, followed by subtype B and the recombinants CRF01_AE and CRF02_AG [14]. The heterogenous distribution of HIV-1 variants is probably the result of founder effects. The greatest genetic diversity of HIV-1 has been found in Africa, consistent with this continent being the source of the epidemic.

In North America, Europe and Australia, subtype B is by far the most common. Therefore, the majority of our knowledge on HIV-1 pathogenesis, diagnosis, antiretroviral treatment and development of antiretroviral drugs is based on subtype B. However, subtype B accounts only for 12% of the total disease burden globally, and various other group M subtypes, and even group O viruses, have been reported in the US and several European countries. HIV-1 subtype C caused globally 47.2% of all new HIV-1 infections [15]. This subtype predominates in all countries in southern Africa, where the AIDS epidemic is explosive, and also in Ethiopia and in India. The second most common clade is subtype A, which caused 30% of all new infections, including CRF01_AE and CRF02_AG [14]. CRF01_AE viruses are responsible for the epidemic in Southeast Asia, and have been documented at low frequencies in several

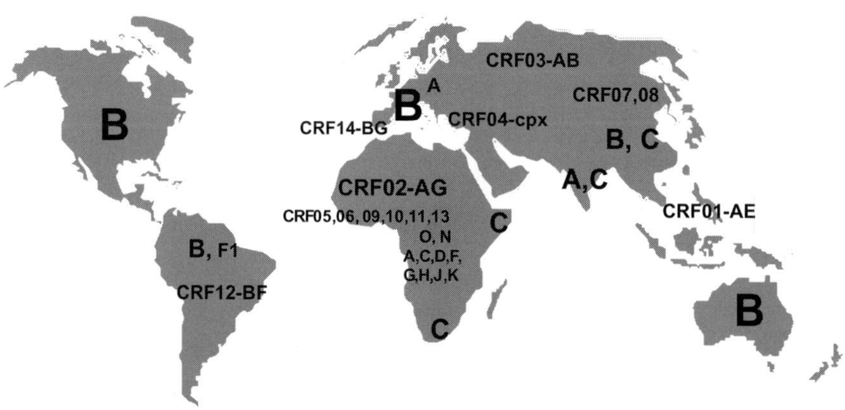

Figure 4. Geographical distribution of HIV worldwide.

Central African countries. CRF02_AG, a complex mosaic of alternating subtype A and G sequences, is the predominant strain in West and West Central Africa. Subtype D is generally limited to Eastern and Central Africa. Subtype F has been reported in Central Africa, South America and Eastern Europe. Subtype G, and A/G recombinant viruses have been observed in Western and Eastern Africa as well in Central Africa. Subtypes H and K have only been detected in Central Africa. Subtype J has been reported in Central Africa and occasionally also in Central America.

Overall, more than 18% of new infections have been attributed to HIV-1 recombinants [10]. In addition to CRF01_AE and CRF02_AG, many other mosaic viruses circulate around the world, but, in contrast, their prevalence seems to be lower, and they often play a major role in certain local epidemics. CRF03_AB is the predominant CRF among intravenous drug users (IDUs) in Kalingrad, in Russia. CRF04_cpx, found in Cyprus and Greece, is a complex mosaic comprising subtypes A, G, H, K and unknown fragments with multiple breakpoints. CRF05_DF has only been identified in Democratic Republic of Congo. CRF06_cpx is a complex mosaic composed of fragments of diverse subtypes A, G, K and J, and circulates principally in West Africa, where it can represent 30–50% of HIV-1 infections in certain countries like Burkina Faso or Niger. CRF07_BC and CRF08_BC are two different inter-subtype B and C recombinants, detected in Northwestern and in Southern China, respectively, mainly among IDUs. CRF09_cpx has been described in several west African countries (Senegal and Ivory Coast), but at low frequencies. CRF10_CD was identified in Tanzania. CRF11-cpx, involving subtypes A, G, J, and CRF01-AE has been detected in Central Africa. CRF12_BF was observed in Argentina and Uruguay. CRF13_cpx is a complex recombinant comprising subtypes A, G, J, and one CRF, CRF01_AE, and was identified in Cameroon. CRF14_BG was documented in IDUs from Spain and Portugal. CRF15_01B found in Thailand is a complex recombinant, comprising CRF and subtypes. CRF16_A2D was identified in Kenya, South Korea and Argentina. At present, 32 different CRFs are described in the HIV Database, but for many of them, no epidemiological background has been reported yet.

It is interesting that many of the CRFs have a restricted geographic spread and are the result of recombination of local co-circulating strains. This illustrates clearly that the global distribution of different forms of HIV-1 is a dynamic process. As more HIV-1 variants inevitably intermix in different parts of the world, the likelihood of generating new recombinant viruses will increase. The pattern of mosaicism will become even more complex, since recombination involving viruses that are already recombinant will also occur (an example being recombination between CRF02 and CRF06 in West Africa). Continued monitoring is necessary to determine the future role of non-subtype B viruses in North America and Europe where they seem to be increasing with time, and to chart the emergence of new predominant subtypes and CRFs around the world.

Challenges of genetic diversity in Env glycoproteins and entry inhibitors

The *pol* gene encodes functional proteins that are targeted by RT inhibitors, and less genetic variation is observed in this gene compared to *env*. Nevertheless, some *in vitro* and *in vivo* observations suggest that genetically distinct viral variants may respond differently to certain antiretroviral drugs that target the *pol* gene. HIV-1 group O and HIV-2 strains are naturally resistant to non-nucleoside RT inhibitors (NNRTIs) [16]. The rate of occurrence of nevirapine resistance-associated mutations after a single dose is significantly higher in women with HIV-1 subtype C than in women with subtype A or D [17]. Many studies have also shown the existence of polymorphisms among non-B strains [18–20]. These accessory (or minor) mutations may not result in a significant decrease in susceptibility [21, 22], but may be associated with an increase in viral fitness (replication capacity) and/or increase in resistance level associated with major mutations, and thus long-term failure of therapy.

In the same way, since entry inhibitors target the highly variable *env* gene, it is likely that antiviral responses of entry inhibitors are even more influenced by the genetic diversity that exists among HIV-1 subtypes and CRFs. In addition to divergence among subtypes, a high intra-patient diversity is also seen; the overall rate of intra-patient divergence of the env gene is close to 1% per year. Slower rates of evolution are found in the other parts of the viral genome. These differences are likely driven by varying selective pressures rather than differences in the underlying mutation rate. Intra-host diversity reflects the successive fixation of advantageous mutations and the extinction of unfavorable lineages [23]. HIV successively fixes mutations that allow it to escape immune responses in the host, or antiviral drugs.

Targeting the envelope glycoproteins (gp41 and gp120) represents a very great challenge because of their high levels of sequence diversity. The entry of HIV into a target cell represents the key initial step in the replication cycle of the virus and involves three different steps: (1) viral attachment of gp120 to CD4 receptor, (2) binding of gp120 to the co-receptors, and (3) fusion of viral and cellular membranes. The entry process thus involves a coordinated series of molecular interactions between the components of the virus glycoprotein complex (gp120/gp41) and the components of the receptor complex (CD4 and a chemokine co-receptor, CXCR4 or CCR5). Upon receptor and co-receptor binding to the surface subunit of gp120, subsequent rearrangements within gp41 allow fusion of viral and cellular membranes. It is important to note that the gp120 compromises five variable (V1–V5) domains interspersed with conserved (C1–C5) regions (see chapter by Tilton/Doms, Fig. 1).

Because of the diversity in the viral glycoproteins and host receptor molecules, their mechanism of action, and consequencely development of resistance, will also differ. Several CD4-gp120 inhibitors are under development. Apparently the degree of sequence variation in the nearby V1/V2 variable regions indirectly influences the susceptibility to these drugs [24]. Therefore, it can be expected that the natural variation in these regions that exist among different subtypes can significantly influence baseline susceptibility to certain of these compounds for some subtypes/CRFs. Moreover, a high intra-patient diversity is also seen in *env* sequences over time.

After binding of the HIV-1 gp120 envelope glycoprotein to CD4, conformational changes occur in the gp120 that translocate the variable regions V1/V2 and V3 of gp120 to create or expose a binding site for co-receptor. An interesting target in HIV entry is the co-receptor binding phase, and current research is focused on designing compounds that interact with the CCR5 and CXCR4 receptors. Resistance to co-receptor antagonists can be the result of either a shift in co-receptor usage or from other changes in the envelope glycoproteins that alter the interaction with co-receptor. Multiple mutations in V3, but also in V2, C2 and V4 seem to account for drug resistance. These regions are known to be highly variable among different subtypes, and, moreover, different co-receptor usage has also been reported for certain subtypes, e.g., CXCR4 variants are rarely observed among subtype C [25], whereas these variants seem to occur at higher frequencies in subtype D [26]. Although preliminary results have not yet identified co-receptor switch as a main resistance pathway, this has to be monitored closely since such shifts could have serious consequences in disease progression. Independent of subtype-related differences, the presence of X4 viruses as minor quasi-species in an individual could allow selection of this variant.

The final step in the viral entry pathway is fusion of the viral envelope with the target surface membrane, and the gp41 ectodomain is the key structure responsible for membrane fusion. Enfuvirtide, previously known as T-20, is the first fusion inhibitor to be approved for the treatment of AIDS. Enfuvirtide is a 36-amino acid peptide derived from a continuous sequence within HR2 from the gp41 from the HXB2 HIV-1 subtype B prototype strain. The peptide binds to the HR1 region and prevents gp41-mediated fusion with the host cell membrane (see Chapter 6 for details). Overall, enfuvirtide should be considered as a drug with a low genetic barrier to resistance [27]. Drug resistance to this new drug seems to occur in the HR1 region and, more precisely, genetic changes within the 36–45 amino acid region of HR1 have been shown to confer resistance to enfuvirtide, especially mutations in the highly conserved 3-amino acid motif at codons 36–38 (GIV) [28]. Other common substitutions observed in phase II and III studies, including Q40H and N42T, and *in vitro* studies showed that site-directed HR1 mutants G36D, V38A, Q40H, N42T, N43D, N43S and N43K are significantly resistant to enfuvirtide. Due to its cutaneous route of administration and its high cost, the use of this drug remains limited to the US and Europe, where subtype B HIV-1 strains predominate. Data on sensitivity and resistance are thus derived mostly from subtype B-infected patients. However, the majority of HIV-1 infections worldwide are with other HIV variants and the proportion of non-B strains is increasing in the western hemisphere. Several studies have examined baseline susceptibility to enfuvirtide using genotypic and/or phenotypic methods. Natural enfuvirtide resistance is rare in B and non-B HIV-1 group M strains [29]. No resistance-associated mutations were seen as natural variants on non-B group M HIV-1 strains. However, other polymorphisms were seen, e.g., Q39L and Q40K, but these mutations were not documented to be associated with resistance, although it has to be further examined to what extent they might affect the accessibility of the drug to its target sequence. On the other hand, the N42S polymorphism has been previously observed in 15% of baseline isolates and is associated with mild hypersusceptibility [30]. This latter mutation is present at the baseline in a large majority of non-B strains, e.g., in 80% of 185 strains studied from Cameroon, and could be detected in almost all the subtypes/CRFs [31]. Analysis of the HR2 domain, from which the peptide is derived, indicated a much greater genetic variability as compared to HR1. Only certain amino acid positions are highly conserved between the different HIV-1 variants and correspond mainly to the amino acids involved in the 5-helix interaction and binding. Despite this high genetic diversity in the HR2 region, the efficacy of enfuvirtide to inhibit replication of such polymorphic strains seems not to be influenced. This was shown by a few studies analyzing the *in vitro* efficacy of T-20 on non-B samples (C, CRF01, CRF02) from Africa and India with more than half of the loci harboring amino acids that are different from the enfuvirtide peptide, suggesting that the HR2-HR1 interaction can tolerate significant genetic changes [32]. Although, the *in vitro* observations together with the highly conserved HR1 regions suggest a broad applicability of T-20 against very diverse HIV-1 group M strains, T-20 is not effective against HIV-2 [30]. Unexpectedly, despite the high genetic diversity in HR1 and HR2, HIV-1 O isolates were as sensitive as group M viruses to inhibition by T-20 *in vitro* and *in vivo* [28, 33]. These findings suggest that T-20 has a broad antiretroviral activity against a large diversity of HIV-1 strains. Other fusion inhibitors targeting HR1 are in clinical development, for example T-1249. Although, group O viruses are susceptible to T-20, polymorphisms in gp41 seem to affect the sensitivity of HIV-1 O to second-generation fusion inhibitors like T-1249. On the other hand, mutations in HR1, known to be associated with resistance to T-20, are not resistant to T-1249. More *in vitro* and *in vivo* studies will be necessary using a larger panel of non-B subtypes to determine the impact of subtypes on enfuvirtide efficacy.

Conclusion

One of the major characteristics of lentiviruses is their extensive genetic variability. This diversity is observed worldwide with a heterogeneous geographical distribution of HIV-1 subtypes and CRFs, but also a high intra-patient diversity (1–5%) is seen. The variability of the *env* gene among the different HIV-1 groups, subtypes worldwide and quasi-species observed within the patient, together with the different regions in the envelope involved (according to which step in viral entry is targeted by the molecule), make it a real challenge to design antiretroviral drugs in this region. Moreover, the genetic diversity of HIV continues to increase, the geographical distribution of subtypes is evolving and intermixing of HIV-1 variants is inevitable. The development of entry inhibitors is mainly based on subtype B sequences and, due to factors related to the complex route of administration or high cost for certain drugs, the clinical evaluation for most entry inhibitors remains limited to the US and Europe, where subtype B HIV-1 strains predominate. Data on susceptibility and development of resistance against entry inhibitors are thus mainly derived from subtype B-infected patients. However, because the majority of HIV-1 infections worldwide involve other subtypes and genetic diversity is increasing, studies of the efficacy of entry inhibitors against non-subtype B viruses will be important from a global perspective.

On January 6th, 2007, Laurence Vergne died tragically in Yaounde, Cameroun. She was only 31 years old. She obtained her PhD degree in 2003 at the University of Montpellier, and had been working with Eric Delaporte and Martine Peeters' team at the Institute for Research and Development (IRD, Institut de Recherche pour le Developpement) in Montpellier, France, since 1998. Her research activities were focused on the genetic diversity of HIV and its implications on antiretroviral drug resistance in developing countries, especially Africa. In Yaounde, Cameroun, with the support of the World Health Organization (WHO-Afro), the International Atomic Energy Agency (IAEA) and ANRS (National Institute for AIDS Research, France), she had just set up a regional reference laboratory for Central Africa for the surveillance and monitoring of genotypic drug resistance of HIV-1. Laurence was a particularly bright young scientist whose generosity towards others had led her to commit herself in the fight against AIDS in Africa. Besides her professional assets, her personal qualities create a terrible sense of loss for all those who knew her.

References

1 Preston BD (1997) Reverse transcriptase and HIV-1 variation. *Science* 275: 228–229

2 Keulen W, Nijhuis M, Schuurman R, Berkhout B, Boucher C (1997) Reverse transcriptase fidelity and HIV-1 variation. *Science* 275: 229–231

3 Hu WS, Temin HM (1990) Retroviral recombination and reverse transcription. *Science* 250: 1227–1233

4 Zhuang J, Jetzt AE, Sun G, Yu H, Klarmann G, Ron Y, Preston BD, Dougherty JP (2002) Human immunodeficiency virus type 1 recombination: rate, fidelity, and putative hot spots. *J Virol* 76: 11273–11282

5 Ho DD, Neuman AU, Perelson AS, Chen W, Leonard JM, Markowitz M (1995) Rapid turnover of plasma virions and CD4 lymphocytes in HIV-1 infection. *Nature* 373: 123–126

6 Coffin JM (1995) HIV population dynamics *in vivo*: implications for genetic variation, pathogenesis, and therapy. *Science* 267: 483–489

7 Perelson AS, Neumann AU, Markowitz M, Leonard JM, Ho DD (1996) HIV-1 dynamics *in vivo*: virion clearance rate, infected cell life-span, and viral generation time. *Science* 271: 1582–1586

8 Robertson DL, Anderson JP, Bradac JA, Carr JK, Foley B, Funkhouser RK, Gao F, Hahn B, Kalish

ML, Kuiken C et al. (2000) HIV-1 nomenclature proposal. *Science* 288: 55–56
9 Schim van der Loeff MF, Aaby P (1999) Towards a better understanding of the epidemiology of HIV-2. *AIDS* 13: S69–84
10 Peeters M, Toure-Kane C, Nkengasong JN (2003) Genetic diversity of HIV in Africa: impact on diagnosis, treatment, vaccine development and trials. *AIDS* 17: 2547–2560
11 Vergne L, Bourgeois A, Mpoudi-Ngolle E, Mougnutou R, Mbuagbaw J, Liegeois F, Laurent C, Butel C, Zekeng L, Delaporte E et al. (2003) Biological and genetic characteristics of HIV infections in Cameroon reveals dual group M and O infections and a correlation between SI-inducing phenotype of the predominant CRF02_AG variant and disease stage. *Virology* 310: 254–266
12 Leitner T, Korber B, Daniels M, Calef C, Foley B (2005) HIV-1 subtype and circulating recombinant form (CRF) reference sequences. In: T Leitner, B Foley, B Hahn, P Marx, F McCutchan, J Mellors, S Wolinsky, B Korber (eds) HIV sequence compendium 2005, Los Alamos National Laboratory, Los Alamos, 41–48
13 Simon F, Mauclere P, Roques P, Loussert-Ajaka I, Muller-Trutwin MC, Saragosti S, Georges-Courbot MC, Barre-Sinoussi F, Brun-Vezinet F (1998) Identification of a new human immunodeficiency virus type 1 distinct from group M and group O. *Nat Med* 4: 1032–1037
14 Osmanov S, Pattou C, Walker N, Schwardlander B, Esparza J; WHO-UNAIDS network for HIV isolation and characterization (2002) Estimated global distribution and incidence of HIV-1 genetic subtypes during year 2000. *J Acquir Immune Defic Syndr* 29: 184–190
15 Julg B, Goebel FD (2005) HIV genetic diversity: any implications for drug resistance? *Infection* 4: 299–301
16 Descamps D, Collin G, Letourneur F, Apetrei C, Damond F, Loussert-Ajaka I, Simon F, Saragosti S, Brun-Vezinet F (1997) Susceptibility of human immunodeficiency virus type 1 group O isolates to antiretroviral agents: *in vitro* phenotypic and genotypic analyses. *J Virol* 71: 8893–8898
17 Eshleman SH, Hoover DR, Chen S, Hudelson SE, Guay LA, Mwatha A, Fiscus SA, Mmiro F, Musoke P, Jackson JB et al. (2005) Nevirapine (NVP) resistance in women with HIV-1 subtype C, compared with subtypes A and D, after the administration of single-dose NVP. *J Infect Dis* 192: 30–36
18 Vergne L, Peeters M, Mpoudi-Ngole E, Bourgeois A, Liegeois F, Toure-Kane C, Mboup S, Mulanga-Kabeya C, Saman E, Jourdan J et al. (2000) Genetic diversity of protease and reverse transcriptase sequences in non-subtype-B human immunodeficiency virus type 1 strains: evidence of many minor drug resistance mutations in treatment-naive patients. *J Clin Microbiol* 38: 3919–3925
19 Pieniazek D, Rayfield M, Hu DJ, Nkengasong J, Wiktor SZ, Downing R, Biryahwaho B, Mastro T, Tanuri A, Soriano V et al. (2000) Protease sequences from HIV-1 group M subtypes A-H reveal distinct amino acid mutation patterns associated with protease resistance in protease inhibitor-naive individuals worldwide. HIV Variant Working Group. *AIDS* 14: 1489–1495
20 Kantor R, Katzenstein D (2003) Polymorphism in HIV-1 non-subtype B protease and reverse transcriptase and its potential impact on drug susceptibility and drug resistance evolution. *AIDS Rev* 5: 25–35
21 Adje-Toure C, Bile CE, Borget MY, Hertog K, Maurice C, Nolan ML, Nkengasong JN (2003) Polymorphism in protease and reverse transcriptase and phenotypic drug resistance of HIV-1 recombinant CRF02_AG isolates from patients with no prior use of antiretroviral drugs in Abidjan, Cote d'Ivoire. *J Acquir Immune Defic Syndr* 34: 111–113
22 Vergne L, Stuyver L, Van Houtte M, Butel C, Delaporte E, Peeters M (2006) Natural polymorphism in protease and reverse transcriptase genes and *in vitro* antiretroviral drug susceptibilities of non-B HIV-1 strains from treatment-naive patients. *J Clin Virol* 36: 43–49
23 Rambaut A, Posada D, Crandall KA, Holmes EC (2004) The causes and consequences of HIV evolution. *Nat Rev Genet* 5: 52–61
24 Madani N, Perdigoto A, Srinivasan K, Cox JM, Chruma JJ, Lalonde J, Head M, Smith AB, Sodroski JG (2004) Localized changes in the gp120 envelope glycoprotein confer resistance to human immunodeficiency virus entry inhibitors BMS-806 and #155. *J Virol* 78: 3742–3752
25 Peeters M, Vincent R, Perret JL, Lasky M, Patrel D, Liegeois F, Courgnaud V, Seng T, Matton, Molinier S et al. (1999) Evidence for differences in MT2 cell tropism according to genetic subtypes of HIV-1: syncytium-inducing variants seem rare among subtype C HIV-1 viruses. *J Acquir Immune Defic Syndr* 20: 115–121
26 Kaleebu P, French N, Mahe C, Yirrell D, Watera C, Lyagoba F, Nakiyingi J, Rutebemberwa A, Morgan D, Weber J et al. (2002) Effect of human immunodeficiency virus (HIV) type 1 envelope

subtypes A and D on disease progression in a large cohort of HIV-1-positive persons in Uganda. *J Infect Dis* 185: 1244–1250
27 Briz V, Poveda E, Soriano V (2006) HIV entry inhibitors: mechanisms of action and resistance pathways. *J Antimicrob Chemother* 57: 619–627
28 Poveda E, Rodes B, Labernardiere JL, Benito JM, Toro C, Gonzalez-Lahoz J, Faudon JL, Clavel F, Schapiro J, Soriano V (2004) Evolution of genotypic and phenotypic resistance to Enfuvirtide in HIV-infected patients experiencing prolonged virologic failure. *J Med Virol* 74: 21–28
29 Roman F, Gonzalez D, Lambert C, Deroo S, Fisher A, Baurith T, Stauub T, Boulme R, Arendt V, Schneider F et al. (2003) Uncommon mutations at residue positions critical for enfuvirtide (T-20) resistance in enfuvirtide-naive patients infected with subtype B and non-B HIV-1 strains. *J Acquir Immune Defic Syndr* 33: 134–139
30 Cooper DA, Lange J (2004) Peptide inhibitors of virus-cell fusion: enfuvirtide as a case study in clinical discovery and development. *Lancet Infect Dis* 4: 426–436
31 Aghokeng AF, Ewane L, Awazi B, Nanfack A, Delaporte E, Zekeng L, Peeters M (2005) Enfuvirtide binding domain is highly conserved in non-B HIV type 1 strains from Cameroon, West Central Africa. *AIDS Res Hum Retroviruses* 21: 430–433
32 Cilliers T, Patience T, Pillay C, Papathanasopoullos M, Morris L (2004) Sensitivity of HIV-1 subtype C isolates to the entry inhibitor T-20. *AIDS Res Hum Retroviruses* 20: 477–482
33 Poveda E, Rodes B, Toro C, Soriano V (2004) Are fusion inhibitors active against all HIV variants? *AIDS Res Hum Retroviruses* 20: 347–348

Entry Inhibitors in HIV Therapy
Edited by Jacqueline D. Reeves and Cynthia A. Derdeyn

Attachment of human immunodeficiency virus to cells and its inhibition

Stefan Pöhlmann[1] and Michel J. Tremblay[2]

[1] *Institute for Clinical and Molecular Virology and Nikolaus-Fiebiger-Center for Molecular Medicine, University Erlangen-Nürnberg, Schlossgarten 4, 91054 Erlangen, Germany*
[2] *Laboratory of Human Immuno-Retrovirology, Research Center in Infectious Diseases, CHUL Research Center, 2705 Laurier Blvd, Quebec, Canada, G1V 4G2, and Faculty of Medicine, Laval University, Quebec, Canada*

Multifaceted events regulate initial interactions between HIV-1 and target cells

The entry of enveloped viruses involves virus adsorption followed by close apposition of the viral and plasma membranes. This multistep process is initiated by specific binding interactions between glycoproteins in the viral envelope and appropriate receptors on the cell surface. In the case of HIV-1, attachment of virions to the cell surface is attributed to a high affinity interaction between envelope spike glycoproteins (Env, composed of the surface protein gp120 and the transmembrane protein gp41) and a complex made of the primary CD4 receptor and a seven-transmembrane co-receptor (e.g., CXCR4 or CCR5) (reviewed in [1]). Then a chain of dynamic events take place that enable the viral nucleocapsid to penetrate within the target cell following the destabilization of membrane microenvironment and the formation of a fusion pore.

Although it is generally accepted that HIV-1 attachment to its major cellular reservoirs (i.e., T helper cells and macrophages) occurs through the two-stage receptor-interaction pathway, there is accumulating evidence indicating that the initial attachment step is a more complex phenomenon than initially thought. Indeed it seems that adsorption of HIV-1 to the cell surface is modulated by a large variety of interactions between the viral entity and the target cell surface (reviewed in [2]). This retrovirus may also attach to some cell types by CD4-independent interactions involving highly glycosylated groups or basic residues found on gp120 and polyanionic sulfated chains or lectin-like domains on some specific cell surface receptors (reviewed in [1]). For example, heparan sulfate proteoglycans, which are expressed at high levels on different cell types, such as epithelial and endothelial cells, can interact with the envelope spike glycoprotein and serve as docking structures for HIV-1 [3]. Heparan sulfate proteoglycans such as syndecans serve as the main class of attachment receptors for HIV-1 on different cell types, e.g., macrophages and endothelial cells, and are thought to play a cardinal role in virus transmission

[4, 5]. GP120 can bind also to galactosyl ceramide and its sulfated derivative (i.e., sulfatide) [6, 7] that are found on macrophages and neural, glial and colon epithelial cells [6–8]. It can also associate with the mannose-specific macrophage endocytosis receptor (MR) [9] and other cellular lectins. In fact, the determinant role played by dendritic cells (DCs) in HIV-1 transmission might rely on specific interactions between gp120 and C-type lectins, of which the DC-specific intercellular adhesion molecule-3 (ICAM-3) grabbing nonintegrin (DC-SIGN) and DC-SIGNR (for DC-SIGN-related) are the best studied [10, 11]. These two lectins are expressed on certain DC populations and endothelial cells, respectively, and are described in more detail in this article.

HIV-1 attachment mediated by host cell proteins incorporated into the viral envelope

Incorporation of host cell surface molecules within nascent HIV-1

HIV-1, as an enveloped virus, is released by budding through the plasma membrane of the productively infected cell. In addition to its own virus-encoded envelope glycoproteins, the virus incorporates many different cellular proteins normally found on the cell surface (reviewed in [12–15]). These include major histocompatibility complex (MHC) class I and II determinants [16–19], adhesion molecules [20–23], complement regulatory proteins [24] and costimulatory molecules [25, 26], which have been found inserted in the viral envelope. The process of incorporation of host cell membrane proteins was found to be conserved among all tested HIV-1 subtypes and strains that were expanded in natural cellular reservoirs, such as mitogen-activated peripheral blood lymphocytes and human lymphoid tissue cultured *ex vivo* [27–32]. The physiological significance of this phenomenon is provided by two previous reports showing that host-encoded cell surface constituents were incorporated in plasma-derived clinical HIV-1 isolates [33, 34]. Although different host cell constituents can be found embedded within HIV-1, the incorporation process seems to be selective. For example, CD45 is the most abundant leukocyte cell surface glycoprotein [35], but is not acquired by HIV-1 [18, 36]. The CXCR4, CCR5, and CCR3 co-receptors are also excluded from HIV-1 [37]. This ability to incorporate discriminatory host antigens into mature virions has allowed two groups to demonstrate that cell-type-specific antigens can serve as markers of the cellular origin of HIV-1 replication [33, 38]. It has been estimated that between 375 and 600 HLA-DR molecules are found associated with HIV-1_{IIIB} emerging from H9 cells [39]. This observation suggests that virally embedded host HLA-DR outnumbered virus envelope (Env) glycoprotein gp120 by a factor of 8.9 to 28.6 considering that HIV-1 possesses an average of between 21 and 42 gp120 molecules per virion [40].

The molecular basis governing the selective incorporation of cell surface proteins within emerging HIV-1 particles is only beginning to be exposed. It was established that the virus envelope spike glycoproteins (i.e., gp120 and gp41) are not essential to achieve insertion of ICAM-1 into HIV-1 [41]. Interestingly, ICAM-1 incorporation is governed by an intimate association between the cytoplasmic domain of ICAM-1 and the viral Gag precursor polyprotein $Pr55^{Gag}$ [42].

Involvement of virus-anchored host proteins and their ligands in the attachment process

It can be proposed that besides interactions between gp120 and multiple attachment receptors, interactions can also occur between host-derived cell surface components incorporated within emerging virions and their natural counter-ligands. This scenario has been confirmed in numerous studies where such host cell membrane molecules were found to retain their biological activity when located on the virus. For example, HLA-DR can increase HIV-1 infectivity for CD4-expressing T cells by about twofold [43], whereas ICAM-1 alone augments virus infectivity for $LFA-1^+$ target cells by up to 100-fold depending of the LFA-1 conformational state [22, 44, 45]. Activation of primary human $CD4^+$ T lymphocytes was found to result in LFA-1 clustering, an event that promotes the early events of HIV-1 replication cycle through an interaction between virus-embedded host ICAM-1 and LFA-1 clusters [46]. Confocal analyses showed that HIV-1 is concentrated in microdomains rich in LFA-1 clusters [46]. Virus entry studies including subcellular fractionation experiments with primary human $CD4^+$ T cells illustrated that the acquisition of ICAM-1 by nascent HIV-1 modified the entry route of the virus within such target cells [47]. It was established that the ICAM-1-mediated increase in virus infectivity was linked with a more productive entry process into primary $CD4^+$ T lymphocytes (i.e., cytosolic delivery of viral material) [47]. It has been reported that the higher susceptibility of memory $CD4^+$ T cells ($CD45RO^+$ subset) to HIV-1 infection is due to secondary interactions between virus-associated ICAM-1 and cell surface LFA-1 [48]. The presence of host-encoded CD28 in newly formed HIV-1 particles resulted in a close to 20-fold augmentation in virus infectivity when using target cells that express high levels of CD80 and CD86, two natural ligands of CD28 [49]. In addition, an increase in virus infectivity was also seen following insertion of host-encoded costimulatory molecules CD80 and CD86 within progeny viruses [50].

Strategies to reduce viral load by blocking interactions between virus-associated host molecules and their physiological counter-receptors

Given that attachment of HIV-1 to host cells can be modulated by the additional interactions provided by virus-anchored host cell membrane proteins, it is thus not surprising to discover that virus susceptibility to blocking agents is affected. For example, ICAM-1-bearing virions are more resistant to antibody-mediated neutralization and this decreased sensitivity is even more dramatic when target cells expressed on their surface the activated form of LFA-1 [51, 52]. Additionally, it was reported that virions carrying host ICAM-1 on their surface are more resistant to the fusion inhibitor T-20 than are isogenic viruses lacking host ICAM-1 [53].

Although the physical presence of such host constituents on the exterior of virions might be detrimental for the infected individual, the propensity of HIV-1 to acquire numerous host cell surface components could be exploited to control viral load. Indeed, it has been shown in numerous reports that HIV-1 infectivity can be efficiently neutralized, both *in vitro* and *in vivo*, with antibodies specific for such host membrane proteins [22, 23, 26, 39, 44, 45, 54, 55]. Interestingly, it was demonstrated that HIV-1 replication is diminished upon treatment with statin compounds (e.g., lovastatin) [56], the primary drugs used in the treatment of hypercholesterolemia. The antiviral potency of lovastatin seems to be linked with its capacity to inhibit interactions between virus-associated host ICAM-1 and cell surface LFA-1. This *in vitro* work was confirmed by a proof-of-concept small-scale clinical study [57]. In this provocative study, six A1 stage HIV-1 patients not receiving combined therapy were given lovastatin for a month as their only medication. This short-term statin treatment clearly reduced serum viral RNA loads in all patients and in general increased their $CD4^+$ T cell counts. Discontinuation of treatment was followed by a rebound in viral load.

HIV-1 attachment mediated by cellular lectins

HIV-1 capture by cellular lectins: Targets for microbicides

The prevention of HIV-1 infection by microbicides, topically applied inhibitors that block access of sexually transmitted HIV-1 to the host system, is an attractive strategy [58]. Understanding which cell types are first targeted by sexually transmitted HIV-1 and how these cells interact with HIV-1 is key to the generation of effective microbicides. Several studies suggest that DCs, professional antigen-presenting cells, might be intimately involved in the early local and subsequent systemic spread of sexually transmitted HIV-1 [59]. Langerhans DCs in the top layer of the anogenital mucosa are probably the first cells exposed to sexually transmitted HIV-1. Mucosal macrophages and submucosal DCs might subsequently get into contact with virus crossing the mucosal barrier via local breaches or with progeny virions generated by infected Langerhans cells. DCs and macrophages express CD4 and chemokine receptors, and are thus permissive to HIV-1 infection, albeit infection of DCs is often relatively inefficient and depends on the subpopulation analyzed [59]. It has been proposed, however, that mere attachment of HIV-1 to mucosa associated DCs might be sufficient to promote HIV-1 spread, since these motile cells might ferry bound virus into lymphoid tissue, the major compartment of HIV-1 replication, as part of their migratory and antigen-presenting functions within the immune system [10]. Several cellular lectins have been implicated in virus attachment to DCs, macrophages and other cell types relevant to HIV-1 spread. Here, we discuss the role of the lectins DC-SIGN, DC-SIGNR, MR and langerin in HIV-1 infection and introduce strategies to inhibit HIV-1 interactions with these molecules.

DC-SIGN and HIV-1: Uptake, processing and MHC presentation versus transmission

DC-SIGN has initially been identified as a gp120-binding calcium-dependent lectin expressed in placental tissue [60]. The lectin has been "rediscovered" in 2000 when Geijtenbeek and colleagues [10] showed that DC-SIGN is expressed on DCs and is involved in HIV-1 binding and subsequent transfer of the virus to T cells, the latter process presumably involving DC-SIGN-dependent endocytosis and conservation of infectious HIV-1 in a low pH compartment [61]. DC-SIGN seemed to mainly account for the ability of DCs to promote HIV-1 infection of cocultured T cells, and it was proposed that DCs might function as Trojan horses, which take up HIV-1 via DC-SIGN and transport the virus into lymphoid tissue [10, 62]. Geijtenbeek and coworkers also provided evidence that DC-SIGN interacts with ICAM-2 on endothelial cells [63] and ICAM-3 on T cells [64], and proposed that these interactions contribute to extravasation of DCs from blood vessels into tissues and to the close contact between DCs and T cells required for efficient antigen presentation, respectively. Thus, a scenario emerged in which DC-SIGN was involved in DC functions critical for the establishment of an effective immune response and simultaneously allowed HIV-1 to misuse DCs to ensure its spread in the host.

A critical contribution of DC-SIGN to DC interactions with T cells/endothelial cells or HIV-1 has subsequently been challenged. It was reported that DC-SIGN or the related protein DC-SIGNR bind to ICAMs with submicromolar affinities similar to that observed for nonspecific cellular proteins [65], suggesting that ICAM recognition might not account for a potential role of DC-SIGN in cell-cell interactions. It was also documented that HIV-1 capture by DCs does either not dependent on DC-SIGN [66, 67] or that the contribution of DC-SIGN is relatively modest with other factors playing an important role [68–71]. In fact, Truville and colleagues [72] provided evidence that different DCs bind to HIV-1 gp120 via different lectins or via CD4, as discussed below. Moreover, it has been demonstrated that transformed cells frequently used to assess DC-SIGN function were not THP-1 monocytes, as reported [10], but most likely Raji B-cells [73], and that these cells as well as monocyte-derived DCs were permissive to infection by HIV-1 [74–76]. The latter observation suggests that the ability of DC-SIGN-expressing cells to maintain HIV-1 infectious over prolonged time is most likely due to productive infection of these cells [74–76]. Indeed, DC-SIGN-dependent HIV-1 transmission is probably a short-lived process (Fig. 1), which is only observed a few hours after the DC-SIGN-positive, HIV-1-exposed cells make contact with target cells. Mainly, HIV-1 might be endocytosed and processed for MHC presentation ([77, 78], Fig. 1). Finally, two reports indicate that DC-SIGN might not be a good marker for DCs *in vivo* [68, 79], with DC-SIGN-positive cells in lymphoid tissue being of macrophage origin [68]. How these results relate to a series of previous studies demonstrating DC-SIGN expression on tissue DCs [10, 80, 81] is currently unclear.

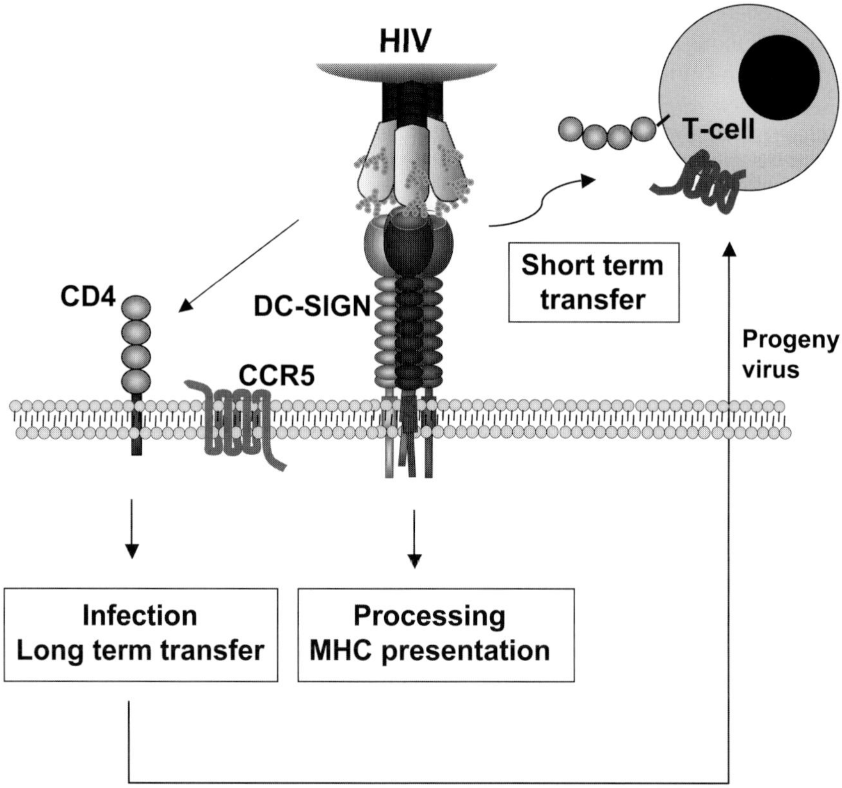

Figure 1. HIV-1 interactions with DC-SIGN on dendritic cells (DCs). DC-SIGN is expressed at high levels on DCs and binds to carbohydrates present on the surface of the heavily glycosylated HIV-1 envelope protein. Binding of HIV-1 to DC-SIGN-positive DCs can have multiple consequences. During a relatively short time window (hours) bound virus can be transferred to adjacent susceptible cells ("short-term transfer"). Certain HIV-1 isolates can also productively infect DCs via CD4 and a chemokine coreceptor. Prior engagement of DC-SIGN might augment infectious entry. Progeny virions produced from infected DCs can then be efficiently transmitted to cocultured T cells over a prolonged time period (days, "long-term transfer"). Finally, HIV-1 captured by DC-SIGN can be endocytosed and processed for MHC presentation.

Can a significant contribution of DC-SIGN to DC interactions with HIV-1, and thus to sexual transmission of HIV-1, be disregarded in the light of these results? Probably not, since several studies also provide evidence for a role of DC-SIGN in HIV-1 capture and transmission by DCs. For example, Arrighi and colleagues [82] demonstrated that siRNA-mediated down-modulation of DC-SIGN diminishes HIV-1 capture by DCs. The contribution of DC-SIGN to this process might be due to an involvement of this lectin in the formation of an infectious synapse [83], a microenvironment established between HIV-1-bearing cells and target cells, which promotes efficient transfer of infectious virions [84]. Interestingly, DC-SIGN did not contribute to HIV-1 infection of target cells in cervical explants but, together with CD4, was mainly responsible for HIV-1 uptake by migratory cells present in these explants [85], suggesting that in HIV-1-infected individuals DC-SIGN might indeed contribute to HIV-1 dissemination by motile cells expressing this lectin. In this regard, it is noteworthy that platelets have been shown to express DC-SIGN and to capture HIV-1 in a largely DC-SIGN-dependent manner [86, 87]. These cells might bind HIV-1 via DC-SIGN once the virus has reached the blood stream and might promote its dissemination in the host system. Similarly, a recent report suggests that a subset of B cells expresses DC-SIGN and facilitates HIV-1 transmission to T cells in a DC-SIGN-dependent manner [88]. Finally, two groups found that certain polymorphisms in the DC-SIGN gene are associated with decreased risk of HIV-1 infection [89, 90], highlighting that DC-SIGN might modulate important events leading to the establishment of HIV-1 infection. Thus, further research is needed to clarify the role of DC-SIGN in HIV-1 infection and to evaluate whether this protein is a potential target for microbicides.

DC-SIGNR polymorphisms and susceptibility to HIV-1 infection

DC-SIGNR [11], also termed L-SIGN (for liver SIGN) [91], shares 77% sequence identity with DC-SIGN and is expressed by sinusoidal endothelial cells in liver (LSECs) and in lymph nodes, alveolar macrophages [92] and enterocytes of the small intestine [93]. Moreover, DC-SIGNR transcripts have been detected at sites of mucosal HIV-1 transmission [94]. DC-SIGNR, like DC-SIGN, binds to high-mannose carbohydrates and captures HIV-1, HIV-2 and simian immunodeficiency virus [11, 91]. Binding to ICAM proteins has also been demonstrated [91]. However, the natural function of DC-SIGNR is currently unclear. Expression of DC-SIGNR in lymph node sinusoids might concentrate HIV-1 in this compartment, while DC-SIGNR on LSECs might promote infection of this cell type, which was shown to be permissive *in vitro* [95] and *in vivo* [96, 97]. LSECs might therefore constantly release progeny virus into the blood stream, thereby promoting HIV-1 spread.

DC-SIGN and DC-SIGNR are both organized into a N-terminal intracellular domain, a transmembrane domain, a neck region containing 7.5 repeats of a 23-amino acid-comprising sequence and a C-terminal lectin domain. In contrast to the neck domain of DC-SIGN, which is highly conserved among individuals, the neck domain of DC-SIGNR is polymorphic. While 7.5 repeats are most often found and are considered wild type (wt), alleles with 5.5 and 6.5 repeats are also frequent (28.9% and 12.2%, respectively, in the Caucasian population [91]). The impact of polymorphisms in the DC-SIGNR neck region on susceptibility to HIV-1 infection has been analyzed by two studies. Lichterfeld and colleagues [98] found no significant differences in DC-SIGNR allele distribution between HIV-1-infected individuals and healthy controls. Also, no correlation between DC-SIGNR allele frequency and course of HIV-1 disease was observed [98]. In contrast, Liu and colleagues [99] found that the 7/7 genotype was significantly less frequent in high-risk HIV-1-seronegative individuals compared to HIV-1-seropositive individuals, while the 5/7 genotype was associated with some protection against HIV-1 infection. It is currently unclear, however, how such a protective effect can be explained on the molecular level. Thus, DC-SIGNR variants with 5 and 6 repeats were found to form stable homo-oligomers [100] and to augment HIV-1 infection [101] with similar efficiency as the wt protein. Also, coexpression of DC-SIGNR alleles with 5 and 7 repeats allowed formation of stable hetero-oligomers and did not result in decreased HIV-1 interactions when compared to controls expressing the 7/7 allele combination [101]. A linkage between DC-SIGNR polymorphisms and alterations in unrelated genes determining susceptibility to HIV-1 infection can therefore at present not be excluded.

MR and langerin mediate HIV-1 gp120 binding to DC subsets

The observation that DCs can bind to HIV-1 independently of DC-SIGN raised the question whether related lectins might be involved. A detailed analysis of gp120 interactions with different DC subsets revealed that MR on dermal DCs might contribute to gp120 capture by these cells [72]. MR is an endocytic receptor that harbors multiple lectin domains and recognizes ligands bearing mannose, fucose or *N*-acetylglucosamine (GlcNac) [102]. The lectin is expressed on DCs, macrophages and some endothelial cells [102] and might contribute to capture of HIV-1 virions by these cells. In fact, it has been demonstrated that an MR-specific antibody can reduce HIV-1 attachment to macrophages [103]. Langerin contains a single carbohydrate recognition domain specific for mannose, fucose and GlcNac and is expressed exclusively by Langerhans cells [104, 105]. Expression of langerin triggers formation of Birbeck granules, which are part of the endosomal recycling machinery of Langerhans cells [106, 107]. The lectin might function as an antigen uptake receptor that releases ligands upon exposure to low pH in endosomal compartments [105]. While Langerin recognizes HIV-1 gp120, it needs to be determined whether it contributes to infection of Langerhans cells, which are susceptible to HIV-1 in culture and in patients [108, 109], or to transmission of HIV-1 from Langerhans cells to adjacent target cells.

Approaches to inhibit HIV-1 interactions with cellular lectins

Lectin-dependent HIV-1 attachment to cells can be prevented by interfering with lectin expression or by targeting domains in the lectin required for efficient ligand recognition. Alternatively, carbohydrate structures in HIV-1-gp120, which are recognized by relevant lectins, are targets for intervention. Down-modulation of lectin expression can be achieved by specific siRNA [82, 110] and by sanglifehrin A [111], an immunosuppressant that diminishes C-type lectin expression on DCs. However, issues with delivery (siRNA) and possible unwanted side effects (sanglifehrin A) need to be addressed. Several inhibitors that impede the interaction of DC-SIGN with HIV-1 or other viruses have been described. A synthetic, branched molecule that presents 32 mannose residues on its surface has been shown to inhibit HIV-1-gp120 binding to DC-SIGN [112] and to block DC-SIGN interactions with reporter viruses bearing the Ebola virus glycoprotein [113], a well-established DC-SIGN ligand [114, 115]. The antiviral activity of comparable molecules bearing sialic acid, the structure recognized by influenza hemagglutinin, has also been demonstrated in a mouse model for influenza infection [116], underlining the feasibility of this approach. The inhibitory substances used to target lectin-mediated HIV-1 attachment must not necessarily be of synthetic origin, since bovine lactoferrin [117] and a substance in human milk which harbors Lewis X carbohydrates [118] were shown to bind to DC-SIGN and to inhibit HIV-1 transmission by DCs. Similarly, a DC-SIGN inhibitory activity was identified in human cervicovaginal lavage fluid [119]. These natural substances might modulate the risk of HIV-1 transmission and merit further investigation. Finally, inhibition of ligand binding to lectins can be achieved by monoclonal antibodies, and a variety of DC-SIGN- or DC-SIGNR-specific monoclonal antibodies that inhibit pathogen interactions with these lectins have been generated [68, 69, 71, 120].

While several lectins expressed at the cell surface can mediate HIV-1 attachment, soluble human-, plant- and bacteria-derived lectins can be employed to inhibit this process. Thus, mannose-binding lectin (MBL), a soluble lectin that is involved in innate immunity and is known to bind to HIV-1-gp120 [121], inhibits DC-SIGN-dependent HIV-1 transmission to target cells, probably by competing with DC-SIGN for binding sites in HIV-1-gp120 [122]. A similar observation was reported for Ebola virus [123], validating that lectins with overlapping carbohydrate specificity can compete for binding sites in gp120, which can result in reduction of viral attachment. In fact, soluble lectins were shown to be effective against HIV-1 transfer by DCs and direct infection of DCs [124], highlighting that lectins applied within a microbicide formulation might help to block HIV-1 infection upon sexual transmission. A promising candidate microbicide is Cyanovirin N, a mannose-specific lectin obtained from the Cyanobacterium *Nostoc ellipsosporum* [125]. CV-N binds to the HIV-1-gp120 protein and inhibits HIV-1 interactions with DCs *in vitro* [124] and, when applied topically, infection of macaques with a simian/human immunodeficiency hybrid virus upon vaginal and rectal challenge [126, 127].

Concluding remarks

A more complete understanding of the possible contribution of virus-associated host proteins to the HIV-1 life cycle is crucial because it might lead to the development of alternative approaches for the treatment of HIV-1 infection and/or the design of an efficient vaccine strategy. Interestingly, a therapeutic or vaccine strategy targeted at virus-associated host cell surface proteins might circumvent problems due to the great genetic variability displayed by HIV-1. Elucidation of the molecular mechanisms underlying HIV-1 capture by cellular lectins and assessment of the contribution of this process to HIV-1 dissemination in and between individuals might help to define novel strategies for preventive or therapeutic intervention. Moreover, lectins on DCs can be used as tools to target HIV-1 antigens to these important antigen-presenting cells [128–131], which might facilitate the generation of an effective HIV-1 vaccine.

Acknowledgements
We acknowledge numerous contributions from various laboratories whose references were not cited in this review due to space limitations. The work presented in this review is supported by grants to M.J.T. from the Canadian Institutes of Health Research HIV/AIDS Research Program (HOP-14438) and by grants from the Deutsche Forschungsgemeinschaft to S.P. (SFB466, GK 1071). M.J.T. is the recipient of the Canada Research Chair in Human Immuno-Retrovirology (senior level).

References

1 Clapham PR, McKnight A (2002) Cell surface receptors, virus entry and tropism of primate lentiviruses. *J Gen Virol* 83: 1809–1829
2 Ugolini S, Mondor I, Sattentau QJ (1999) HIV-1 attachment: another look. *Trends Microbiol* 7: 144–149
3 Mondor I, Ugolini S, Sattentau QJ (1998) Human immunodeficiency virus type 1 attachment to HeLa CD4 cells is CD4 independent and gp120 dependent and requires cell surface heparans. *J Virol* 72: 3623–3634
4 Saphire AC, Bobardt MD, Zhang Z, David G, Gallay PA (2001) Syndecans serve as attachment receptors for human immunodeficiency virus type 1 on macrophages. *J Virol* 75: 9187–9200
5 Bobardt MD, Saphire AC, Hung HC, Yu X, Van der Schueren B, Zhang Z, David G, Gallay PA (2003) Syndecan captures, protects, and transmits HIV to T lymphocytes. *Immunity* 18: 27–39
6 Fantini J, Cook DG, Nathanson N, Spitalnik SL, Gonzalez-Scarano F (1993) Infection of colonic epithelial cell lines by type 1 human immunodeficiency virus is associated with cell surface expression of galactosylceramide, a potential alternative gp120 receptor. *Proc Natl Acad Sci USA* 90: 2700–2704
7 Harouse JM, Bhat S, Spitalnik SL, Laughlin M, Stefano K, Silberberg DH, Gonzalez-Scarano F (1991) Inhibition of entry of HIV-1 in neural cell lines by antibodies against galactosyl ceramide. *Science* 253: 320–323

8 Seddiki N, Ramdani A, Saffar L, Portoukalian J, Gluckman JC, Gattegno L (1994) A monoclonal antibody directed to sulfatide inhibits the binding of human immunodeficiency virus type 1 (HIV-1) envelope glycoprotein to macrophages but not their infection by the virus. *Biochim Biophys Acta* 1225: 289–296
9 Larkin M, Childs RA, Matthews TJ, Thiel S, Mizuochi T, Lawson AM, Savill JS, Haslett C, Diaz R, Feizi T (1989) Oligosaccharide-mediated interactions of the envelope glycoprotein gp120 of HIV-1 that are independent of CD4 recognition. *AIDS* 3: 793–798
10 Geijtenbeek TB, Kwon DS, Torensma R, van Vliet SJ, van Duijnhoven GC, Middel J, Cornelissen IL, Nottet HS, KewalRamani VN, Littman DR et al. (2000) DC-SIGN, a dendritic cell-specific HIV-1-binding protein that enhances trans-infection of T cells. *Cell* 100: 587–597
11 Pöhlmann S, Soilleux EJ, Baribaud F, Leslie GJ, Morris LS, Trowsdale J, Lee B, Coleman N, Doms RW (2001) DC-SIGNR, a DC-SIGN homologue expressed in endothelial cells, binds to human and simian immunodeficiency viruses and activates infection in trans. *Proc Natl Acad Sci USA* 98: 2670–2675
12 Tremblay MJ, Fortin J-F, Cantin R (1998) The acquisition of host-encoded proteins by nascent HIV-1. *Immunol Today* 19: 346–351
13 Ott DE (2002) Potential roles of cellular proteins in HIV-1. *Rev Med Virol* 12: 359–374
14 Ott DE (1997) Cellular proteins in HIV virions. *Rev Med Virol* 7: 167–180
15 Cantin R, Methot S, Tremblay MJ (2005) Plunder and stowaways: incorporation of cellular proteins by enveloped viruses. *J Virol* 79: 6577–6587
16 Cantin R, Fortin J-F, Lamontagne G, Tremblay M (1997) The acquisition of host major histocompatibility complex class II glycoproteins by human immunodeficiency virus type 1 accelerates the process of virus entry and infection in human T-lymphoid cells. *Blood* 90: 1091–1100
17 Cantin R, Martin G, Tremblay MJ (2001) A novel virus capture assay reveals a differential acquisition of host HLA-DR by clinical isolates of human immunodeficiency virus type 1 expanded in primary human cells depending on the nature of producing cells and the donor source. *J Gen Virol* 82: 2979–2987
18 Orentas RJ, Hildreth JEK (1993) Association of host cell surface adhesion receptors and other membrane proteins with HIV and SIV. *AIDS Res Hum Retroviruses* 9: 1157–1165
19 Rossio JL, Bess J, Henderson LE, Cresswell P, Arthur LO (1995) HLA class II on HIV particles is functional in superantigen presentation to human T cells: Implications for HIV pathogenesis. *AIDS Res Hum Retroviruses* 11: 1433–1439
20 Liao Z, Roos JW, Hildreth JE (2000) Increased infectivity of HIV type 1 particles bound to cell surface and solid-phase ICAM-1 and VCAM-1 through acquired adhesion molecules LFA-1 and VLA-4. *AIDS Res Hum Retroviruses* 16: 355–366
21 Guo MML, Hildreth JEK (1995) HIV acquires functional adhesion receptors from host cells. *AIDS Res Hum Retroviruses* 11: 1007–1013
22 Fortin J-F, Cantin R, Tremblay M (1998) T cells expressing activated LFA-1 are more susceptible to infection with human immunodeficiency virus type 1 particles bearing host-encoded ICAM-1. *J Virol* 72: 2105–2112
23 Rizzuto CD, Sodroski JG (1997) Contribution of virion ICAM-1 to human immunodeficiency virus infectivity and sensitivity to neutralization. *J Virol* 71: 4847–4851
24 Montefiori DC, Cornell RJ, Zhou JY, Zhou JT, Hirsch VM, Johnson PR (1994) Complement control proteins, CD46, CD55, and CD59, as common surface constituents of human and simian immunodeficiency viruses and possible targets for vaccine protection. *Virology* 205: 82–92
25 Giguère J-F, Paquette JS, Bounou S, Cantin R, Tremblay MJ (2002) New insights into the functionality of a virion-anchored host cell membrane protein: CD28 *versus* HIV type 1. *J Immunol* 169: 2762–2771
26 Bounou S, Dumais N, Tremblay MJ (2001) Attachment of human immunodeficiency virus-1 (HIV-1) particles bearing host-encoded B7-2 proteins leads to nuclear factor-κB- and nuclear factor of activated T cells-dependent activation of HIV-1 long terminal repeat transcription. *J Biol Chem* 276: 6359–6369
27 Cantin R, Fortin J-F, Tremblay M (1996) The amount of host HLA-DR proteins acquired by HIV-1 is virus strain- and cell type-specific. *Virology* 218: 372–381
28 Capobianchi MR, Fais S, Castilletti C, Gentile M, Ameglio F, Dianzani F (1994) A simple and reliable method to detect cell membrane proteins on infectious human immunodeficiency virus type 1 particles. *J Infect Dis* 169: 886–889
29 Frank I, Stoiber H, Godar S, Stockinger H, Steindl F, Katinger HWD, Dierich MP (1996)

Acquisition of host cell-surface-derived molecules by HIV-1. *AIDS* 10: 1611–1620
30 Roberts BD, Butera ST (1999) Host protein incorporation is conserved among diverse HIV-1 subtypes. *AIDS* 13: 425–427
31 Bounou S, Leclerc JE, Tremblay MJ (2002) Presence of host ICAM-1 in laboratory and clinical strains of human immunodeficiency virus type 1 increases virus infectivity and CD4(+)-T- cell depletion in human lymphoid tissue, a major site of replication *in vivo*. *J Virol* 76: 1004–1014
32 Bastiani L, Laal S, Kim M, Zolla-Pazner S (1997) Host cell-dependent alterations in envelope components of human immunodeficiency virus type 1 virions. *J Virol* 71: 3444–3450
33 Lawn SD, Roberts BD, Griffin GE, Folks TM, Butera ST (2000) Cellular compartments of human immunodeficiency virus type 1 replication *in vivo*: determination by presence of virion-associated host proteins and impact of opportunistic infection. *J Virol* 74: 139–145
34 Saarloos M-N, Sullivan BL, Czerniewski MA, Parameswar KD, Spear GT (1997) Detection of HLA-DR associated with monocytotropic, primary, and plasma isolates of human immunodeficiency virus type 1. *J Virol* 71: 1640–1643
35 Trowbridge IS (1991) CD45: A prototype for transmembrane protein tyrosine phosphatases. *J Biol Chem* 26: 23517–23520
36 Esser MT, Graham DR, Coren LV, Trubey CM, Bess JW, Arthur LO, Ott DE, Lifson JD (2001) Differential incorporation of CD45, CD80 (B7-1), CD86 (B7-2), and major histocompatibility complex class I and II molecules into human immunodeficiency virus type 1 virions and microvesicles: implications for viral pathogenesis and immune regulation. *J Virol* 75: 6173–6182
37 Lallos LB, Laal S, Hoxie JA, Zolla-Pazner S, Bandres JC (1999) Exclusion of HIV coreceptors CXCR4, CCR5, and CCR3 from the HIV envelope. *AIDS Res Hum Retroviruses* 15: 895–897
38 Frank I, Kacani L, Stoiber H, Stossel H, Spruth M, Steindl F, Romani N, Dierich MP (1999) Human immunodeficiency virus type 1 derived from cocultures of immature dendritic cells with autologous T cells carries T-cell-specific molecules on its surface and is highly infectious. *J Virol* 73: 3449–3454
39 Arthur LO, Bess JWJ, Sowder II RC, Benveniste RE, Mann DL, Cherman J-C, Henderson LE (1992) Cellular proteins bound to immunodeficiency viruses: Implication for pathogenesis and vaccines. *Science* 258: 1935–1938
40 Chertova E, Bess JW Jr, Crise BJ, Sowder IR, Schaden TM, Hilburn JM, Hoxie JA, Benveniste RE, Lifson JD, Henderson LE et al. (2002) Envelope glycoprotein incorporation, not shedding of surface envelope glycoprotein (gp120/SU), Is the primary determinant of SU content of purified human immunodeficiency virus type 1 and simian immunodeficiency virus. *J Virol* 76: 5315–5325
41 Beauséjour Y, Tremblay MJ (2004) Envelope glycoproteins are not required for insertion of host ICAM-1 into human immunodeficiency virus type 1 and ICAM-1-bearing viruses are still infectious despite a suboptimal level of trimeric envelope proteins. *Virology* 324: 165–172
42 Beauséjour Y, Tremblay MJ (2004) Interaction between the cytoplasmic domain of ICAM-1 and Pr55Gag leads to acquisition of host ICAM-1 by human immunodeficiency virus type 1. *J Virol* 78: 11916–11925
43 Cantin R, Fortin J-F, Lamontagne G, Tremblay M (1997) The presence of host-derived HLA-DR1 on human immunodeficiency virus type 1 increases viral infectivity. *J Virol* 71: 1922–1930
44 Fortin J-F, Cantin R, Lamontagne G, Tremblay M (1997) Host-derived ICAM-1 glycoproteins incorporated on human immunodeficiency virus type 1 are biologically active and enhances viral infectivity. *J Virol* 71: 3588–3596
45 Fortin JF, Barbeau B, Hedman H, Lundgren E, Tremblay MJ (1999) Role of the leukocyte function antigen-1 conformational state in the process of human immunodeficiency virus type 1-mediated syncytium formation and virus infection. *Virology* 257: 228–238
46 Tardif MR, Tremblay MJ (2005) Regulation of LFA-1 activity through cytoskeleton remodeling and signaling components modulates the efficiency of HIV type-1 entry in activated $CD4^+$ T lymphocytes. *J Immunol* 175: 926–935
47 Tardif MR, Tremblay MJ (2003) Presence of host ICAM-1 in human immunodeficiency virus type 1 virions increases productive infection of $CD4^+$ T lymphocytes by favoring cytosolic delivery of viral material. *J Virol* 77: 12299–12309
48 Tardif MR, Tremblay MJ (2005) LFA-1 is a key determinant for preferential infection of memory $CD4^+$ T cells by human immunodeficiency virus type 1. *J Virol* 79: 13714–13724
49 Giguère JF, Paquette J-S, Bounou S, Cantin R, Tremblay MJ (2002) New insights into the functionality of a virion-anchored host cell membrane protein: CD28 vs HIV type 1. *J Immunol* 169: 2762–2771

50 Giguère J-F, Bounou S, Paquette JS, Madrenas J, Tremblay MJ (2004) Insertion of host-derived costimulatory molecules CD80 (B7.1) and CD86 (B7.2) into human immunodeficiency virus type 1 affects the virus life cycle. *J Virol* 78: 6222–6232
51 Fortin J-F, Cantin R, Bergeron MG, Tremblay MJ (2000) Interaction between virion-bound host ICAM-1 and the high affinity state of LFA-1 on target cells renders R5 and X4 isolates of HIV-1 more refractory to neutralization. *Virology* 268: 493–503
52 Losier M, Fortin JF, Cantin R, Bergeron MG, Tremblay MJ (2003) Virion-bound ICAM-1 and activated LFA-1: a combination of factors conferring resistance to neutralization by sera from human immunodeficiency virus type 1-infected individuals independently of the disease status and phase. *Clin Immunol* 108: 111–118
53 Beauséjour Y, Tremblay MJ (2004) Susceptibility of HIV type 1 to the fusion inhibitor T-20 is reduced on insertion of host intercellular adhesion molecule 1 in the virus membrane. *J Infect Dis* 190: 894–902
54 Arthur LO, Bess JWJ, Urban RG, Strominger JL, Morton WR, Mann DL, Henderson LE, Benveniste RE (1995) Macaques immunized with HLA-DR are protected from challenge with simian immunodeficiency virus. *J Virol* 69: 3117–3124
55 Chan WL, Rodgers A, Grief C, Almond N, Ellis S, Flanagan B, Silvera P, Bootman J, Stott J, Kent K et al. (1995) Immunization with class I histocompatibility leukocyte antigen can protect macaques against challenge infection with SIVmac-32H. *AIDS* 9: 223–228
56 Giguère J-F, Tremblay MJ (2004) Statin compounds reduce human immunodeficiency virus type 1 replication by preventing the interaction between virion-associated host intercellular adhesion molecule 1 and its natural cell surface ligand LFA-1. *J Virol* 78: 12062–12065
57 del Real G, Jimenez-Baranda S, Mira E, Lacalle RA, Lucas P, Gomez-Mouton C, Alegret M, Pena JM, Rodriguez-Zapata M, Alvarez-Mon M et al. (2004) Statins inhibit HIV-1 infection by down-regulating Rho activity. *J Exp Med* 200: 541–547
58 Shattock RJ, Moore JP (2003) Inhibiting sexual transmission of HIV-1 infection. *Nat Rev Microbiol* 1: 25–34
59 Steinman RM, Granelli-Piperno A, Pope M, Trumpfheller C, Ignatius R, Arrode G, Racz P, Tenner-Racz K (2003) The interaction of immunodeficiency viruses with dendritic cells. *Curr Top Microbiol Immunol* 276: 1–30
60 Curtis BM, Scharnowske S, Watson AJ (1992) Sequence and expression of a membrane-associated C-type lectin that exhibits CD4-independent binding of human immunodeficiency virus envelope glycoprotein gp120. *Proc Natl Acad Sci USA* 89: 8356–8360
61 Kwon DS, Gregorio G, Bitton N, Hendrickson WA, Littman DR (2002) DC-SIGN-mediated internalization of HIV is required for trans-enhancement of T cell infection. *Immunity* 16: 135–144
62 Geijtenbeek TB, Engering A, Van Kooyk Y (2002) DC-SIGN, a C-type lectin on dendritic cells that unveils many aspects of dendritic cell biology. *J Leukoc Biol* 71: 921–931
63 Geijtenbeek TB, Krooshoop DJ, Bleijs DA, van Vliet SJ, van Duijnhoven GC, Grabovsky V, Alon R, Figdor CG, van Kooyk Y (2000) DC-SIGN-ICAM-2 interaction mediates dendritic cell trafficking. *Nat Immunol* 1: 353–357
64 Geijtenbeek TB, Torensma R, van Vliet SJ, van Duijnhoven GC, Adema GJ, van Kooyk Y, Figdor CG (2000) Identification of DC-SIGN, a novel dendritic cell-specific ICAM-3 receptor that supports primary immune responses. *Cell* 100: 575–585
65 Snyder GA, Ford J, Torabi-Parizi P, Arthos JA, Schuck P, Colonna M, Sun PD (2005) Characterization of DC-SIGN/R interaction with human immunodeficiency virus type 1 gp120 and ICAM molecules favors the receptor's role as an antigen-capturing rather than an adhesion receptor. *J Virol* 79: 4589–4598
66 Wu L, Bashirova AA, Martin TD, Villamide L, Mehlhop E, Chertov AO, Unutmaz D, Pope M, Carrington M, KewalRamani VN (2002) Rhesus macaque dendritic cells efficiently transmit primate lentiviruses independently of DC-SIGN. *Proc Natl Acad Sci USA* 99: 1568–1573
67 Gummuluru S, Rogel M, Stamatatos L, Emerman M (2003) Binding of human immunodeficiency virus type 1 to immature dendritic cells can occur independently of DC-SIGN and mannose binding C-type lectin receptors via a cholesterol-dependent pathway. *J Virol* 77: 12865–12874
68 Granelli-Piperno A, Pritsker A, Pack M, Shimeliovich I, Arrighi JF, Park CG, Trumpfheller C, Piguet V, Moran TM, Steinman RM (2005) Dendritic cell-specific intercellular adhesion molecule 3-grabbing nonintegrin/CD209 is abundant on macrophages in the normal human lymph node and is not required for dendritic cell stimulation of the mixed leukocyte reaction. *J Immunol* 175: 4265–4273

69 Baribaud F, Pöhlmann S, Leslie G, Mortari F, Doms RW (2002) Quantitative expression and virus transmission analysis of DC-SIGN on monocyte-derived dendritic cells. *J Virol* 76: 9135–9142
70 Trumpfheller C, Park CG, Finke J, Steinman RM, Granelli-Piperno A (2003) Cell type-dependent retention and transmission of HIV-1 by DC-SIGN. *Int Immunol* 15: 289–298
71 Wu L, Martin TD, Vazeux R, Unutmaz D, KewalRamani VN (2002) Functional evaluation of DC-SIGN monoclonal antibodies reveals DC-SIGN interactions with ICAM-3 do not promote human immunodeficiency virus type 1 transmission. *J Virol* 76: 5905–5914
72 Turville SG, Cameron PU, Handley A, Lin G, Pöhlmann S, Doms RW, Cunningham AL (2002) Diversity of receptors binding HIV on dendritic cell subsets. *Nat Immunol* 3: 975–983
73 Wu L, Martin TD, Carrington M, KewalRamani VN (2004) Raji B cells, misidentified as THP-1 cells, stimulate DC-SIGN-mediated HIV transmission. *Virology* 318: 17–23
74 Turville SG, Santos JJ, Frank I, Cameron PU, Wilkinson J, Miranda-Saksena M, Dable J, Stossel H, Romani N, Piatak M Jr et al. (2004) Immunodeficiency virus uptake, turnover, and 2-phase transfer in human dendritic cells. *Blood* 103: 2170–2179
75 Nobile C, Petit C, Moris A, Skrabal K, Abastado JP, Mammano F, Schwartz O (2005) Covert human immunodeficiency virus replication in dendritic cells and in DC-SIGN-expressing cells promotes long-term transmission to lymphocytes. *J Virol* 79: 5386–5399
76 Burleigh L, Lozach PY, Schiffer C, Staropoli I, Pezo V, Porrot F, Canque B, Virelizier JL, Arenzana-Seisdedos F, Amara A (2006) Infection of dendritic cells (DCs), not DC-SIGN-mediated internalization of human immunodeficiency virus, is required for long-term transfer of virus to T cells. *J Virol* 80: 2949–2957
77 Moris A, Nobile C, Buseyne F, Porrot F, Abastado JP, Schwartz O (2004) DC-SIGN promotes exogenous MHC-I-restricted HIV-1 antigen presentation. *Blood* 103: 2648–2654
78 Moris A, Pajot A, Blanchet F, Guivel-Benhassine F, Salcedo M, Schwartz O (2006) Dendritic cells and HIV-specific $CD4^+$ T cells: HIV antigen presentation, T cell activation, viral transfer. *Blood* 108: 1643–1651
79 Krutzik SR, Tan B, Li H, Ochoa MT, Liu PT, Sharfstein SE, Graeber TG, Sieling PA, Liu YJ, Rea TH et al. (2005) TLR activation triggers the rapid differentiation of monocytes into macrophages and dendritic cells. *Nat Med* 11: 653–660
80 Jameson B, Baribaud F, Pöhlmann S, Ghavimi D, Mortari F, Doms RW, Iwasaki A (2002) Expression of DC-SIGN by dendritic cells of intestinal and genital mucosae in humans and rhesus macaques. *J Virol* 76: 1866–1875
81 Soilleux EJ, Morris LS, Leslie G, Chehimi J, Luo Q, Levroney E, Trowsdale J, Montaner LJ, Doms RW, Weissman D et al. (2002) Constitutive and induced expression of DC-SIGN on dendritic cell and macrophage subpopulations *in situ* and *in vitro*. *J Leukoc Biol* 71: 445–457
82 Arrighi JF, Pion M, Wiznerowicz M, Geijtenbeek TB, Garcia E, Abraham S, Leuba F, Dutoit V, Ducrey-Rundquist O, van Kooyk Y et al. (2004) Lentivirus-mediated RNA interference of DC-SIGN expression inhibits human immunodeficiency virus transmission from dendritic cells to T cells. *J Virol* 78: 10848–10855
83 Arrighi JF, Pion M, Garcia E, Escola JM, van Kooyk Y, Geijtenbeek TB, Piguet V (2004) DC-SIGN-mediated infectious synapse formation enhances X4 HIV-1 transmission from dendritic cells to T cells. *J Exp Med* 200: 1279–1288
84 McDonald D, Wu L, Bohks SM, KewalRamani VN, Unutmaz D, Hope TJ (2003) Recruitment of HIV and its receptors to dendritic cell-T cell junctions. *Science* 300: 1295–1297
85 Hu Q, Frank I, Williams V, Santos JJ, Watts P, Griffin GE, Moore JP, Pope M, Shattock RJ (2004) Blockade of attachment and fusion receptors inhibits HIV-1 infection of human cervical tissue. *J Exp Med* 199: 1065–1075
86 Boukour S, Masse JM, Benit L, Dubart-Kupperschmitt A, Cramer EM (2006) Lentivirus degradation and DC-SIGN expression by human platelets and megakaryocytes. *J Thromb Haemost* 4: 426–435
87 Chaipan C, Soilleux EJ, Simpson P, Hofmann H, Gramberg T, Marzi A, Geier M, Stewart EA, Eisemann J, Steinkasserer A et al. (2006) DC-SIGN and CLEC-2 mediate human immunodeficiency virus type 1 capture by platelets. *J Virol* 80:8951–60
88 Rappocciolo G, Piazza P, Fuller CL, Reinhart TA, Watkins SC, Rowe DT, Jais M, Gupta P, Rinaldo CR (2006) DC-SIGN on B lymphocytes is required for transmission of HIV-1 to T lymphocytes. *PLoS Pathog* 2: e70
89 Martin MP, Lederman MM, Hutcheson HB, Goedert JJ, Nelson GW, van Kooyk Y, Detels R, Buchbinder S, Hoots K, Vlahov D et al. (2004) Association of DC-SIGN promoter polymorphism

with increased risk for parenteral, but not mucosal, acquisition of human immunodeficiency virus type 1 infection. *J Virol* 78: 14053–14056
90 Liu H, Hwangbo Y, Holte S, Lee J, Wang C, Kaupp N, Zhu H, Celum C, Corey L, McElrath MJ et al. (2004) Analysis of genetic polymorphisms in CCR5, CCR2, stromal cell-derived factor-1, RANTES, and dendritic cell-specific intercellular adhesion molecule-3-grabbing nonintegrin in seronegative individuals repeatedly exposed to HIV-1. *J Infect Dis* 190: 1055–1058
91 Bashirova AA, Geijtenbeek TB, van Duijnhoven GC, van Vliet SJ, Eilering JB, Martin MP, Wu L, Martin TD, Viebig N, Knolle PA et al. (2001) A dendritic cell-specific intercellular adhesion molecule 3-grabbing nonintegrin (DC-SIGN)-related protein is highly expressed on human liver sinusoidal endothelial cells and promotes HIV-1 infection. *J Exp Med* 193: 671–678
92 Jeffers SA, Tusell SM, Gillim-Ross L, Hemmila EM, Achenbach JE, Babcock GJ, Thomas WD Jr, Thackray LB, Young MD, Mason RJ et al. (2004) CD209L (L-SIGN) is a receptor for severe acute respiratory syndrome coronavirus. *Proc Natl Acad Sci USA* 101: 15748–15753
93 Chan VS, Chan KY, Chen Y, Poon LL, Cheung AN, Zheng B, Chan KH, Mak W, Ngan HY, Xu X et al. (2006) Homozygous L-SIGN (CLEC4M) plays a protective role in SARS coronavirus infection. *Nat Genet* 38: 38–46
94 Liu H, Hladik F, Andrus T, Sakchalathorn P, Lentz GM, Fialkow MF, Corey L, McElrath MJ, Zhu T (2005) Most DC-SIGNR transcripts at mucosal HIV transmission sites are alternatively spliced isoforms. *Eur J Hum Genet* 13: 707–715
95 Steffan AM, Lafon ME, Gendrault JL, Schweitzer C, Royer C, Jaeck D, Arnaud JP, Schmitt MP, Aubertin AM, Kirn A (1992) Primary cultures of endothelial cells from the human liver sinusoid are permissive for human immunodeficiency virus type 1. *Proc Natl Acad Sci USA* 89: 1582–1586
96 Housset C, Lamas E, Brechot C (1990) Detection of HIV1 RNA and p24 antigen in HIV1-infected human liver. *Res Virol* 141: 153–159
97 Housset C, Lamas E, Courgnaud V, Boucher O, Girard PM, Marche C, Brechot C (1993) Presence of HIV-1 in human parenchymal and non-parenchymal liver cells *in vivo*. *J Hepatol* 19: 252–258
98 Lichterfeld M, Nischalke HD, van Lunzen J, Sohne J, Schmeisser N, Woitas R, Sauerbruch T, Rockstroh JK, Spengler U (2003) The tandem-repeat polymorphism of the DC-SIGNR gene does not affect the susceptibility to HIV infection and the progression to AIDS. *Clin Immunol* 107: 55–59
99 Liu H, Carrington M, Wang C, Holte S, Lee J, Greene B, Hladik F, Koelle DM, Wald A, Kurosawa K et al. (2006) Repeat-region polymorphisms in the gene for the dendritic cell-specific intercellular adhesion molecule-3-grabbing nonintegrin-related molecule: effects on HIV-1 susceptibility. *J Infect Dis* 193: 698–702
100 Guo Y, Atkinson CE, Taylor ME, Drickamer K (2006) All but the shortest polymorphic forms of the viral receptor DC-SIGNR assemble into stable homo- and heterotetramers. *J Biol Chem* 281: 16794–16798
101 Gramberg T, Zhu T, Chaipan C, Marzi A, Liu H, Wegele A, Andrus T, Hofmann H, Pöhlmann S (2006) Impact of polymorphisms in the DC-SIGNR neck domain on the interaction with pathogens. *Virology* 347: 354–363
102 Gordon S (2002) Pattern recognition receptors: doubling up for the innate immune response. *Cell* 111: 927–930
103 Nguyen DG, Hildreth JE (2003) Involvement of macrophage mannose receptor in the binding and transmission of HIV by macrophages. *Eur J Immunol* 33: 483–493
104 Valladeau J, Duvert-Frances V, Pin JJ, Dezutter-Dambuyant C, Vincent C, Massacrier C, Vincent J, Yoneda K, Banchereau J, Caux C et al. (1999) The monoclonal antibody DCGM4 recognizes Langerin, a protein specific of Langerhans cells, and is rapidly internalized from the cell surface. *Eur J Immunol* 29: 2695–2704
105 Stambach NS, Taylor ME (2003) Characterization of carbohydrate recognition by langerin, a C-type lectin of Langerhans cells. *Glycobiology* 13: 401–410
106 McDermott R, Bausinger H, Fricker D, Spehner D, Proamer F, Lipsker D, Cazenave JP, Goud B, De La Salle H, Salamero J et al. (2004) Reproduction of Langerin/CD207 traffic and Birbeck granule formation in a human cell line model. *J Invest Dermatol* 123: 72–77
107 Valladeau J, Ravel O, Dezutter-Dambuyant C, Moore K, Kleijmeer M, Liu Y, Duvert-Frances V, Vincent C, Schmitt D, Davoust J et al. (2000) Langerin, a novel C-type lectin specific to Langerhans cells, is an endocytic receptor that induces the formation of Birbeck granules.

Immunity 12: 71–81
108 Tschachler E, Groh V, Popovic M, Mann DL, Konrad K, Safai B, Eron L, diMarzo Veronese F, Wolff K, Stingl G (1987) Epidermal Langerhans cells – a target for HTLV-III/LAV infection. *J Invest Dermatol* 88: 233–237
109 Rappersberger K, Gartner S, Schenk P, Stingl G, Groh V, Tschachler E, Mann DL, Wolff K, Konrad K, Popovic M (1988) Langerhans' cells are an actual site of HIV-1 replication. *Intervirology* 29: 185–194
110 Nair MP, Reynolds JL, Mahajan SD, Schwartz SA, Aalinkeel R, Bindukumar B, Sykes D (2005) RNAi-directed inhibition of DC-SIGN by dendritic cells: prospects for HIV-1 therapy. *AAPS J* 7: E572–578
111 Woltman AM, Schlagwein N, van der Kooij SW, van Kooten C (2004) The novel cyclophilin-binding drug sanglifehrin A specifically affects antigen uptake receptor expression and endocytic capacity of human dendritic cells. *J Immunol* 172: 6482–6489
112 Tabarani G, Reina JJ, Ebel C, Vives C, Lortat-Jacob H, Rojo J, Fieschi F (2006) Mannose hyperbranched dendritic polymers interact with clustered organization of DC-SIGN and inhibit gp120 binding. *FEBS Lett* 580: 2402–2408
113 Lasala F, Arce E, Otero JR, Rojo J, Delgado R (2003) Mannosyl glycodendritic structure inhibits DC-SIGN-mediated Ebola virus infection in cis and in trans. *Antimicrob Agents Chemother* 47: 3970–3972
114 Simmons G, Reeves JD, Grogan CC, Vandenberghe LH, Baribaud F, Whitbeck JC, Burke E, Buchmeier MJ, Soilleux EJ, Riley JL et al. (2003) DC-SIGN and DC-SIGNR bind ebola glycoproteins and enhance infection of macrophages and endothelial cells. *Virology* 305: 115–123
115 Alvarez CP, Lasala F, Carrillo J, Muniz O, Corbi AL, Delgado R (2002) C-type lectins DC-SIGN and L-SIGN mediate cellular entry by Ebola virus in cis and in trans. *J Virol* 76: 6841–6844
116 Landers JJ, Cao Z, Lee I, Piehler LT, Myc PP, Myc A, Hamouda T, Galecki AT, Baker JR Jr (2002) Prevention of influenza pneumonitis by sialic acid-conjugated dendritic polymers. *J Infect Dis* 186: 1222–1230
117 Groot F, Geijtenbeek TB, Sanders RW, Baldwin CE, Sanchez-Hernandez M, Floris R, van Kooyk Y, de Jong EC, Berkhout B (2005) Lactoferrin prevents dendritic cell-mediated human immunodeficiency virus type 1 transmission by blocking the DC-SIGN–gp120 interaction. *J Virol* 79: 3009–3015
118 Naarding MA, Ludwig IS, Groot F, Berkhout B, Geijtenbeek TB, Pollakis G, Paxton WA (2005) Lewis X component in human milk binds DC-SIGN and inhibits HIV-1 transfer to $CD4^+$ T lymphocytes. *J Clin Invest* 115: 3256–3264
119 Jendrysik MA, Ghassemi M, Graham PJ, Boksa LA, Williamson PR, Novak RM (2005) Human cervicovaginal lavage fluid contains an inhibitor of HIV binding to dendritic cell-specific intercellular adhesion molecule 3-grabbing nonintegrin. *J Infect Dis* 192: 630–639
120 Dakappagari N, Maruyama T, Renshaw M, Tacken P, Figdor C, Torensma R, Wild MA, Wu D, Bowdish K, Kretz-Rommel A (2006) Internalizing antibodies to the C-type lectins, L-SIGN and DC-SIGN, inhibit viral glycoprotein binding and deliver antigen to human dendritic cells for the induction of T cell responses. *J Immunol* 176: 426–440
121 Ji X, Gewurz H, Spear GT (2005) Mannose binding lectin (MBL) and HIV. *Mol Immunol* 42: 145–152
122 Spear GT, Zariffard MR, Xin J, Saifuddin M (2003) Inhibition of DC-SIGN-mediated trans infection of T cells by mannose-binding lectin. *Immunology* 110: 80–85
123 Ji X, Olinger GG, Aris S, Chen Y, Gewurz H, Spear GT (2005) Mannose-binding lectin binds to Ebola and Marburg envelope glycoproteins, resulting in blocking of virus interaction with DC-SIGN and complement-mediated virus neutralization. *J Gen Virol* 86: 2535–2542
124 Turville SG, Vermeire K, Balzarini J, Schols D (2005) Sugar-binding proteins potently inhibit dendritic cell human immunodeficiency virus type 1 (HIV-1) infection and dendritic-cell-directed HIV-1 transfer. *J Virol* 79: 13519–13527
125 Barrientos LG, Gronenborn AM (2005) The highly specific carbohydrate-binding protein cyanovirin-N: structure, anti-HIV/Ebola activity and possibilities for therapy. *Mini Rev Med Chem* 5: 21–31
126 Tsai CC, Emau P, Jiang Y, Agy MB, Shattock RJ, Schmidt A, Morton WR, Gustafson KR, Boyd MR (2004) Cyanovirin-N inhibits AIDS virus infections in vaginal transmission models. *AIDS Res Hum Retroviruses* 20: 11–18
127 Tsai CC, Emau P, Jiang Y, Tian B, Morton WR, Gustafson KR, Boyd MR (2003) Cyanovirin-N

gel as a topical microbicide prevents rectal transmission of SHIV89.6P in macaques. *AIDS Res Hum Retroviruses* 19: 535–541

128 Gieseler RK, Marquitan G, Hahn MJ, Perdon LA, Driessen WH, Sullivan SM, Scolaro MJ (2004) DC-SIGN-specific liposomal targeting and selective intracellular compound delivery to human myeloid dendritic cells: implications for HIV disease. *Scand J Immunol* 59: 415–424

129 Korokhov N, de Gruijl TD, Aldrich WA, Triozzi PL, Banerjee PT, Gillies SD, Curiel TJ, Douglas JT, Scheper RJ, Curiel DT (2005) High efficiency transduction of dendritic cells by adenoviral vectors targeted to DC-SIGN. *Cancer Biol Ther* 4: 289–294

130 Tacken PJ, de Vries IJ, Gijzen K, Joosten B, Wu D, Rother RP, Faas SJ, Punt CJ, Torensma R, Adema GJ et al. (2005) Effective induction of naive and recall T-cell responses by targeting antigen to human dendritic cells via a humanized anti-DC-SIGN antibody. *Blood* 106: 1278–1285

131 Maguire CA, Sapinoro R, Girgis N, Rodriguez-Colon SM, Ramirez SH, Williams J, Dewhurst S (2006) Recombinant adenovirus type 5 vectors that target DC-SIGN, ChemR23 and alpha(v)beta3 integrin efficiently transduce human dendritic cells and enhance presentation of vectored antigens. *Vaccine* 24: 671–682

Edited by Jacqueline D. Reeves and Cynthia A. Derdeyn

Inhibitors that target gp120-CD4 interactions

Pin-fang Lin[1], John Kadow[2] and Louis Alexander[1]

Departments of Virology[1] and Chemistry[2] Bristol-Myers Squibb Pharmaceutical Research Institute, 5 Research Parkway, Wallingford, CT 06492, USA

Introduction

The treatment of HIV-1 is moving towards chronic management of the disease, e.g., by combining three-drug regimens to reduce the number of dosing units. However, four classes (nucleoside/nucleotide and non-nucleoside reverse transcriptase, protease, and fusion inhibitors) including 24 approved drugs are still inadequate and treatment failures continue to occur. Factors contributing to such failures include: the emergence of drug-resistant strains, suboptimal exposure, and poor adherence that is mainly attributable to side effects. Moreover, the transmission of drug-resistant viruses is expected to rise over time. For these reasons, there is a pressing need for new classes of antiretroviral agents that are effective against HIV-1 resistant or insensitive to current drugs and that have the potential for co-formulation in convenient dosing regimens.

HIV-1 entry into host cells, which begins with the binding of the envelope protein gp120 to cellular CD4 receptors, is an attractive target for the development of novel antiretroviral agents. CD4 is primarily expressed on the surface of T cells and macrophages. The protein consists of an extracellular region of 370 amino acid residues organized into four domains (D1–D4), with the HIV-1-binding site of CD4 being localized to D1. In particular, the primary binding site for gp120 maps to positions F43 and R59 on CD4 [1]. Sequence analysis of HIV-1 isolates has identified five highly glycosylated variable (V1–V5) regions of gp120 interspersed with five conserved regions (C1–C5). Intramolecular disulfide bonding in V1–V4 results in loop formation, and C1–C5 fold to form the gp120 core. This structural design allows genetically diverse HIV-1 isolates to retain a common mechanism for cell entry using the conserved core, while simultaneously evading the host immune system [2]. These concepts are based, in part, on the elucidated crystal structure of the HIV-1 gp120 core (lacking the V1–V2, V3 loops, and the N and C termini) in complex with a two-domain fragment of CD4 and the Fab fragment of 17b, a monoclonal antibody (mAb) that recognizes the CCR5 co-receptor-binding region of gp120 [1]. In this CD4-bound structure, the inner core domain and the variable outer domain as well as a bridging sheet contribute to interactions with CD4. More recently, the structure of a fully glycosylated simian immunodeficiency virus (SIV) gp120 core (excluding the V1–V2, V3 loops, and parts of the N and C termini) in an unliganded conformation was solved [3]. Comparison of the structure of unliganded gp120 with that of the CD4-bound protein demonstrates that the inner domain undergoes extensive conformational rearrangement upon receptor binding to assemble the binding pocket for CD4. The opening of this conserved recessed pocket is occupied by the F43 residue of CD4 [1]. The CD4-bound gp120 structure reveals direct interatomic contacts between 22 CD4 residues, and 26 gp120 amino acid residues that are distributed over the entire length of gp120. Most importantly, the F43 of CD4 makes multiple contacts centered on residues E370, I371, N425, M426, W427, G473, and D368 of gp120, which accounts for 23% of the total interatomic contacts in gp120/CD4 interaction. This conserved CD4-binding pocket of gp120 provides a potential target for compound binding.

The observations that the CD4-IgG fusion protein PRO542 (Progenic) and a small-molecule CD4-competition inhibitor BMS-488043 (Bristol-Myers Squibb, BMS) have shown significant clinical efficacy provide encouraging data to support the contention that targeting the gp120/CD4 interaction has therapeutic potential [4, 5]. This chapter discusses the profiles, developmental status, issues, and potential of HIV-1 inhibitors of gp120-CD4 interaction. The efficacy of this class of inhibitors as microbicides in monkeys for preventing viral transmission is presented in the chapter by Clyde Hart.

Inhibitors in development

Targeting the CD4-binding pocket of gp120

Small-molecule inhibitors

BMS-378806 and BMS-488043: BMS-378806 and BMS-488043 (BMS) (Tab. 1) are novel, small-molecule CD4-attachment inhibitors that specifically target gp120 [6–9]. The compounds exhibit potent antiviral activity (nM range) against macrophage-, T-, and dual-tropic HIV-1 laboratory strains (B subtype), indicating that their antiviral activities are coreceptor independent [6]. However, they exhibit a range of activities against HIV-1 strains of other subtypes, likely due to the heterogeneous nature of gp120 sequences. As expected, susceptibility is retained against variant HIV-1 viruses resistant to other classes of antiretroviral agents, and these compounds show no significant cytotoxicity at concentrations $\leq$225 μM.

BMS-378806 and BMS-488043 block HIV-1 entry by a similar mechanism; selectively binding to gp120 with an 1:1 stoichiometry and competitively inhibiting the interaction of soluble CD4 (sCD4) with gp120 [8, 9]. Mutually exclusive binding of BMS-488043 and sCD4 to gp120 is observed using either labeled sCD4 or compound in the presence of an unlabeled ligand, and prior binding of sCD4 to gp120 negates BMS-488043 inhibition. Similarly, this

Table 1. Profiles of selected gp120-CD4 inhibitors

Name	Structure	Target site	Mechanisam	Development status	Source	Ref.
BMS-378806	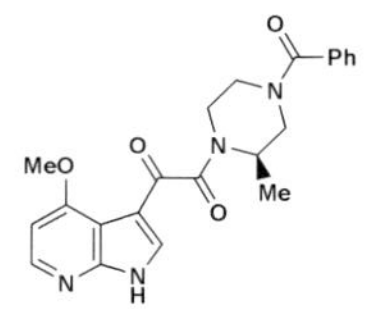	CD4 binding site on gp120	Blocks CD4-bindng to gp120	Phase I discontinued	Bristol-Myers Squibb	[6, 7]
BMS-488043	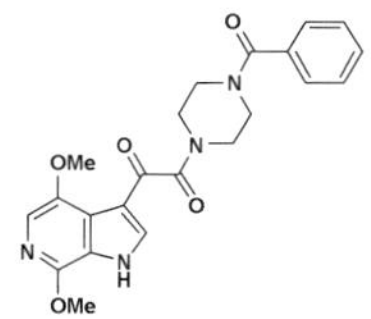	CD4 binding site on gp120	Blocks CD4-bindng to gp120	Phase IIa supplanted by superior compounds	Bristol-Myers Squibb	[9]
NBD-556 (R = Cl) NBD-557 (R = Br)	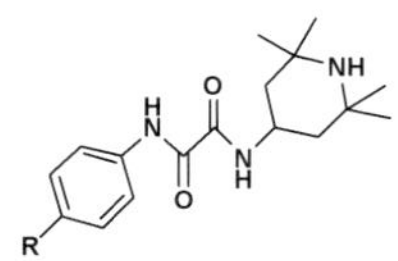	gp120	Blocks CD4-gp120 interactions	Preclinical	New York Blood Center	[12]
WO 2005/016344 Example 12	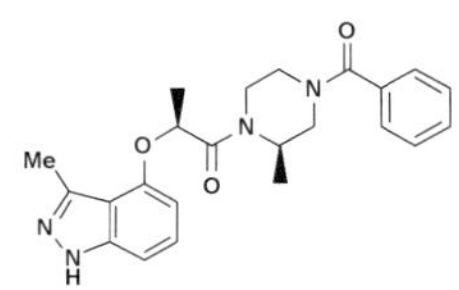	Not specified, likely gp120	Blocks CD4-gp120 interactions	Preclinical	Pfizer	[13]

(Continued on next page)

Table 1. (Continued)

Name	Structure	Target site	Mechanisam	Development status	Source	Ref.
WO 2005/121094 Example 18	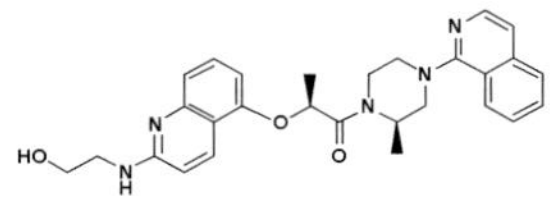	Not specified, likely gp120	Blocks CD4-gp120 interactions	Preclinical	Pfizer	[14]
WO2005/121175 Compound 5	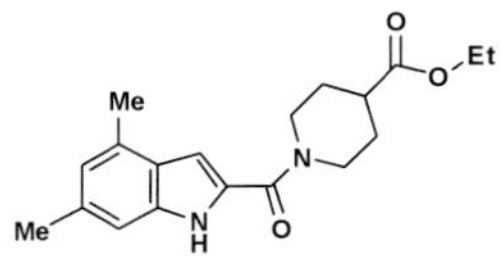	CD4 mimetic binds to gp120	Induces conformational change in envelope to expose cryptic epitopes for antibodies	Preclinical	Chiron	[15]
Bivalent inhibitor	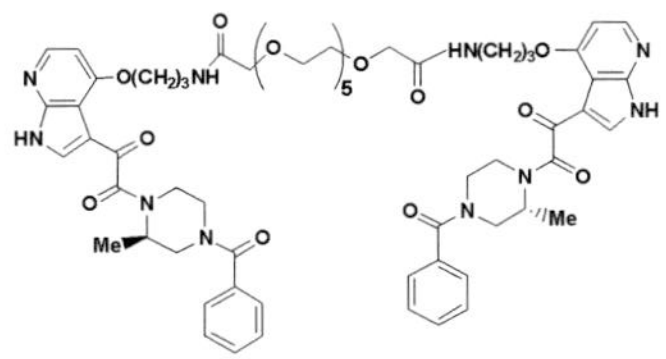	CD4 binding site on gp120	Not specified	Preclinical	Univ. of Maryland	[16]
Pro 542	sCD4-IgG2 Ab	CD4 binding site on gp120	Blocks CD4-gp120 interactions	Phase II	Progenics	[5, 17, 18]
sCD4-17b	D1D2 of sCD4 linked to 17b monoclonal Ab	CD4 and coreceptor binding sites on gp120	Blocks CD4 and co-coreceptor interactions with gp120	Preclinical	NIH	[19]
CD4 M33	27 amino acid CD4 mimic	CD4 binding site on gp120	Blocks CD4-gp120 interactions	Preclinical	CEA Saclay	[21]

(Continued on next page)

Table 1. (Continued)

Name	Structure	Target site	Mechanisam	Development status	Source	Ref.
IgG1b12	mAb against gp120	CD4 binding site on gp120	Blocks CD4-gp120 interactions	Preclinical	Scripps Research Institute	[25, 26]
CADA	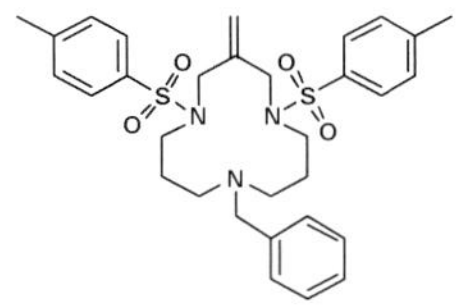	CD4 related pathways	CD4 down modulation	Preclinical	Rega Institute	[27]
NSC-13778	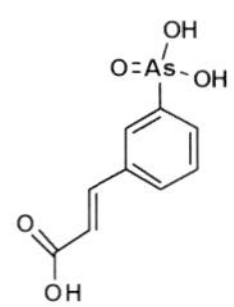	D1/D2 of CD4	Block at least part of the gp120-binding site on CD4	Preclinical	NCI	[28]
TNX-355	Humanized mAb against CD4	D2 of CD4	Prevent CD4-induced conformational changes in gp120	Phase II	Tanox	[30, 32]

compound prevents virion envelope trimers from binding to sCD4 (sCD4-IgG) and blocks sCD4-induced exposure of the gp41 groove in virions [9]. In these virion-binding assays, BMS-488043 is inactive when added subsequent to sCD4 addition. Together, these results indicate that interference of gp120/sCD4 interactions is the primary inhibition mechanism of this compound. BMS-488043 appears to interfere with these interactions by inducing conformational changes at both the CD4- and CCR5-binding regions of gp120 [9], likely inhibiting the acquisition of key structure(s) required for CD4 binding [1].

An alternative inhibition mechanism has been proposed for these BMS compounds in which HIV-1 entry is prevented by blocking CD4-induced gp120 conformational changes without significantly affecting CD4 binding [10]. However, further investigations suggest that the lack of observed sCD4 inhibition in this study is likely due to the high concentrations of sCD4 used in the gp120/CD4 binding assay [9]. Moreover, the observed lack of inhibition of sCD4 binding to a cell surface-expressed envelope variant containing a specific C-terminal deletion is likely caused by a significantly altered envelope structure. This deletion is known to affect gp120 conformation resulting in greater exposure of the CD4- and coreceptor-binding regions. Consequently, BMS-488043 and BMS-378806 only partially inhibit sCD4 binding to this laboratory-generated mutant HIV-1 virion. The effective inhibition of its corresponding pseudotype virus infection may result from an additional mechanism [9]. Thus, the preponderance of experimental evidence indicates that the BMS inhibitors block viral entry by altering envelope conformation and by primarily interfering with CD4 binding.

In cell culture, the rate of resistance development to BMS-378806 is comparable to that observed for nevirapine and lamivudine. The major resistance substitutions selected by this compound in HIV_{NL4-3} and HIV_{LAI} span the CD4-binding pocket, as well as the entire gp120 and gp41 regions, confirming that the viral Envelope is the target of inhibitor [6, 8]. Studies using *in vitro* mutagenized Envelope demonstrate that the changes M426L and M475I confer high levels of resistance in recombinant virus and gp120/CD4-binding assays, and cause a ≥100-fold decrease in BMS-378806 binding to gp120. These changes are located at or near the gp120/CD4 contact sites, supporting the premise that the CD4-binding pocket is the target for this compound [6, 8]. Moreover, a W427V substitution in gp120 fully negates both BMS-378806 and sCD4 binding. This important Envelope change resulted in a nonviable virus, and thus, was not identified in drug selection experiments. In addition, the S375W mutant envelope in which a tryptophan residue occupies the CD4-F43-binding pocket does not bind to BMS-378806 [8]. Together, the data suggest that BMS-378806 interacts with a selective subset of gp120 residues, and that the compound binding site is situated in the CD4-F43-binding pocket of gp120. This premise is supported by studies of the effect of *in vitro* mutagenized gp120-$HXCB_c2$ HIV variants on drug susceptibility, in which the alteration of five CD4-F43 cavity contacting residues of gp120 affect BMS-378806 susceptibility. Other gp120 changes associated with BMS-378806 resistance line a water-filled channel that extends from the F43 cavity to the gp120 surface [11]. In further support of these observations, the potential BMS-378806-binding site was recently revealed by mapping BMS-378806 resistance substitutions to a deep, hydrophobic channel to which many CD4 contact sites also mapped in the X-ray crystal model of unliganded SIV gp120 [3]. Interestingly, some BMS-378806-selected secondary substitutions are also located in the CCR5-binding sites or the gp41 region, suggesting that resistance may result from the interactions of various regions of the HIV-1 Envelope [6, 8, 11]. This result is not surprising because these inhibitors act by inducing Envelope structural modulations [9]. Collectively, resistance mapping, SIV gp120 structure modeling, and biochemical data indicate that BMS inhibitors and sCD4 likely share some common binding sites on the unliganded form of gp120, and CD4-binding inhibition is the primary mechanism of this series of inhibitors.

Phase I studies of BMS-378806 revealed an acceptable safety profile, but its development was terminated due to its failure to achieve target exposure levels. Subsequently, BMS-488043 was advanced to efficacy studies after the Phase I trials of single and multiple oral doses in healthy subjects demonstrated adequate safety parameters, tolerability and pharmacokinetic profiles for up to 14 days. The antiviral activity, safety, and tolerability of BMS-488043 were evaluated in a multiple-dose study in HIV-infected adults that either received 800- or 1800-mg doses of BMS-488043 or placebo every 12 h for 8 days. Whereas 67% of the compound-treated patients had a viral load decline ranging from >0.5 to 1.5 $\log_{10}$, none of placebo-treated subjects had a maximal viral load decline >0.5 $\log_{10}$ copies/mL. Moreover, there were no serious adverse events; hence, monotherapy with BMS-488043 for 8 days was generally safe and well-tolerated [4]. Although BMS-488043 has now been supplanted in development by compounds with improved preclinical profiles, the results of these clinical studies provide proof of concept for this series of compounds, i.e., an orally bioavailable small-molecule CD4-attachment inhibitor of HIV-1 can have potent antiviral activity in infected subjects.

NBD-556 and NBD-557: Both NBD-556 and NBD-557 (The New York Blood Center) are small-molecule inhibitors [*N*-phenyl-*N'*-(2,2,6,6-tetramethyl-piperidin-4-yl)-oxalamide analogs, (Tab. 1)] that block gp120-CD4 interactions [12]. They bind to unliganded gp120 but not to the cellular CD4 receptor in a surface plasmon resonance study and have no activity against HIV-1 reverse transcriptase, protease, or integrase. Moreover, NBD-556 and NBD-557 exhibit micromolar potency against selected HIV-1 laboratory strains and possess minimal cytotoxicity. Both compounds display similar efficacy against laboratory isolates and a limited number of primary HIV-1 strains of other subtypes. This activity is coreceptor independent and does not interfere with gp41 six-helix bundle formation in an *in vitro* model system. However, NBD-related compounds are ~1000-fold less potent than BMS-378806, although their low molecular weight and simple, drug-like structures suggests potential for further optimization.

Pfizer compounds: Pfizer has disclosed, in two published patent applications, a series of small molecules that inhibit the interactions of gp120 with CD4. Example 12 from the initial application (Tab. 1) displayed an IC_{50} of 15 nM in a cell-cell fusion assay and an IC_{50} of 750 nM in a gp120/CD4-binding assay [13]. In a second application, a compound (Example 18, Tab. 1) displayed an IC_{50} of 40 pM in a fusion assay [14]. Although these initial data are encouraging, additional profiling and virology data will be needed to assess the potential of these inhibitors as new clinical candidates.

Chiron CD4 mimetics: A published patent application from Chiron describes small-molecule CD4 mimetics that bind to HIV-1 Envelope proteins, cause conformational changes, and induce the exposure of cryptic epitopes [15]. For example, a complex of compound 5 from the patent application (Tab. 1) and gp120 exhibited enhanced binding to antibody 17b in comparison to CD4-envelope complexes, suggesting that it competes with CD4 and increases binding of the Envelope to 17b.

University of Maryland inhibitors: In addition to the development of molecularly distinct compounds, attempts have been made to alter BMS-378806 in the hopes of improving its preclinical profile. A synthetic bivalent inhibitor (Tab. 1) composed of two BMS-378806 molecules tethered via the C-4 positions is essentially equipotent to BMS-378806, and much more potent than a 3-amino propyl ether monomer precursor version of this compound [16]. Thus, the dimer appears to regain the potency lost by the monovalent modification. Furthermore, structure/activity relationship studies on primarily the C-4 position of both indoles and 7-azaindoles were described.

CD4 IgG fusion protein Pro-542

Pro-542 (Progenics) is a fusion protein comprised of human IgG2 in which the Fv portions of both heavy and light chains have been replaced with four copies of the envelope-binding region of the CD4 receptor (domains 1 and 2). The protein binds to gp120 and blocks entry of HIV-1 into $CD4^+$ T cells. Due to its increased valency and conformational flexibility, Pro-542 broadly and potently neutralizes diverse primary HIV-1 isolates [17]. Moreover, in early clinical trials, PRO542 demonstrated a serum half life of >2 days and significant antiviral activity without exhibiting appreciable toxicity at doses of up to 10 mg/kg in HIV-infected patients [18]. In a Phase II single-dose study at 25 mg/kg, PRO542 mediated an 80% response rate and statistically significant ~0.5 log mean reductions in viral load for 4–6 weeks post treatment. In addition, a significant correlation was observed between the antiviral effect observed *in vivo* and viral susceptibility to PRO542 *in vitro* [5]. Despite these promising characteristics, intravenous administration and drug cost may be obstacles to wider clinical use. PRO542 is also being evaluated for its potential as a microbicide (see Chapter 7).

Chimeric protein sCD4-17b

sCD4-17b is a fusion protein that contains D1/D2 of CD4 attached via a flexible polypeptide linker to a single chain of the variable region of the human mAb 17b that binds to the co-receptor-binding region of gp120. In culture, sCD4-17b neutralizes a broad spectrum of HIV-1 isolates from subtypes A–F that utilize either the CCR5 or CXCR4 co-receptors, and exhibits better inhibitory activity than mAbs IgGb12, 2G12 (targeting the carbohydrate moieties of gp120), and 2F5 (a gp41 mAb). However, several primary isolates are insensitive to this inhibitor [19]. It has been postulated that passive immunization with sCD4-17b could be effective in blocking productive HIV-1 infection (including use as a topical microbicide), although its potency as well as its side effects in humans subjected to long-term administration need to be determined.

CD4 mimetic peptides

An initial 33-residue peptide inhibitor was generated by replacing a loop of the scorpion toxin charybdtoxin with an equivalent of the gp120-binding CDR2-like loop from CD4, which interacts directly with gp120. This miniprotein competitively inhibited gp120-CD4 interactions and elicited antibodies specifically recognizing CD4. A derivative of this miniprotein also induced exposure of CD4i epitopes. Optimization efforts resulted in an enhanced mini-CD4 protein (CD4M9, 28mer) that inhibited replication of both HIV-1 laboratory-adapted and primary isolates, albeit with ~100-fold lower potency than that of a native sCD4 [20]. Further engineering of CD4M9 with non-natural amino acid substitutions led to an optimized 27 amino acid CD4 mimetic (CD4M33) that inhibited the binding of CD4 to gp120 at nanomolar concentrations and blocked infection by primary HIV-1 isolates [21]. Although these later versions display promising potency, issues associated with the development of this compound class include molecular size, pharmacokinetic properties and drug delivery.

To target multiple CD4-binding pockets in trimeric gp120, bivalent CD4 miniprotein inhibitors that contained two CD4M9 moieties, tethered by a spacer of varied length, were synthesized and evaluated [22]. The synthetic bivalent miniproteins showed 5–21-fold enhancement in anti-HIV activity over the monovalent miniprotein with the most potent molecule displaying an EC_{50} of 120 nM. In another approach, a small-molecule (MW 810), water-soluble, proteolytically stable CD4 β-turn mimetic (residues Q40–T45 of the gp120-binding region of CD4) displayed a low μM *K*d for gp120 and reduced syncytium formation. A patent application covering CD4 designed mimics has also been published [23]. Furthermore, a phage display approach modifying the charabdotoxin scaffold resulted in the identification of effective gp120 binders lacking F43 [24]. This method is potentially useful for assessing

CD4/gp120 binding structural determinants and may facilitate the design of new attachment inhibitors. However, an alternative attempt to design small-molecule analogs to mimic the crucial features of CD4-F43 and -R59 residues was unsuccessful.

Neutralizing mAb IgG1 b12

The well-characterized neutralizing mAb IgG1 b12 (b12) that blocks gp120-CD4 interactions is capable of inhibiting a variety of subtype B strains, and prevents infection of other subtypes, albeit with decreased potency. The broadness of b12 anti-HIV activity is related to its capacity to bind to a very conserved region of gp120. Specifically, b12 uses its elongated heavy chain third hypervariable loop to obstruct an essential Trp residue in the CD4-binding pocket [25, 26]. Although the aforementioned characteristics of b12 make it an attractive potential antiretroviral agent, its utility may be limited by its production cost and issues associated with delivery of a protein drug.

Targeting the CD4 receptor

CADA: down-regulation of CD4

The macrocycle cyclotriazadisulfonamide CADA (Tab. 1) has specific interactions with CD4-related pathways that lead to a decrease in cell surface and intracellular CD4, but does not affect the surface expression of other molecules. The CD4-down-modulation activity of CADA analogs directly correlates with their anti-HIV activity, and the activity of this class of compounds is reversible. CADA is effective against a wide range of R5 and X4 utilizing HIV-1 isolates at micromolar ranges. Interestingly, selected resistance mutations (S438R and S463P) map to the CD4-binding region of gp120, and the compound shows cross-resistance to anti-CD4 mAbs. CADA is in the preclinical stage of development and its long-term tolerability must be determined before clinical use can be attempted [27].

NSC 13778: Binding to gp120-binding domain of CD4

NSC 13778 is a pentavalent antimony compound (MW 319, Tab. 1) initially identified via a cell-based anti-HIV-1 screen [28]. The compound binds to the D1–D2 domain of CD4, and appears to compete with gp120 for CD4 binding. NSC 13778 blocks the infection of X4- and R5-tropic HIV-1 strains into $CD4^+$ T cells with micromolar potency, and exhibits minimal cytotoxicity. Moreover, this compound does not reduce the lymphoproliferative response to tetanus toxoid or phytohemagglutinin, suggesting that it might not interfere with immune responses, at least in the short-term. T cells incubated with NSC 13778 show decreased reactivity to anti-CD4 mAbs known to recognize the gp120-binding site. Thus, NSC 13778 may block at least part of the gp120-binding site on CD4, and thereby prevent the recognition by gp120 on incoming HIV-1 virions. NSC 13778 or its analogs are also inhibitors of B-zip transcription factor/DNA binding [29]. Although the discovery of the ability of NSC 13778 to inhibit HIV-1 entry is an interesting finding, additional research will be needed to assess the specificity and suitability of these stibonic acid class compounds for use as long-term oral drugs.

Monoclonal antibody: TNX-355

TNX-355, a humanized IgG4 mAb, blocks HIV-1 entry into host cells by binding to domain 2 of CD4. Unlike antibodies that target domain 1 of CD4, TNX-355 does not interfere with antigen presentation. It is also unlikely to induce $CD4^+$ T cell depletion and complement mediated cytotoxicity. Binding of TNX-355 to CD4 does not compete with binding to gp120. Instead, it functions by preventing CD4-induced conformational changes that are required for HIV-1 entry. TNX-355 inhibits diverse HIV-1 isolates with EC_{50} values ranging from 0.4 to 152 ng/ml [30, 31]. The clinical effectiveness of TNX-355 has been demonstrated in humans following single-dose administrations of up to 25 mg/kg, a dose at which a reduction of plasma HIV-1 RNA was sustained for 28 days and $CD4^+$ T cell increases were observed [32]. A 48-week Phase II study demonstrated antiviral activity and increases in $CD4^+$ T cell count when TNX-355 was administered in combination with an optimized background regimen. Further clinical development is ongoing.

Conclusions

Inhibition of gp120/CD4 interactions represents an attractive target for the development of new antiretroviral therapies and topical microbicides since inhibition of HIV-1 entry is the first line of defense against viral infection. Compounds that prevent virus entry have the potential to: (1) be efficacious in patients already infected with viruses resistant to marketed drugs, (2) reduce the number of latent reservoirs, (3) decrease gp120-mediated cytopathic effects, (4) boost immune clearance of virus by altering the conformation of the viral Envelope and inducing exposure of cryptic epitopes, and/or (5) be included in new therapeutic combinations, as indicated by the synergistic antiviral effects observed *in vitro* by combination of two HIV-1 entry inhibitors [33, 34].

Additional work will be needed to assess which of the aforementioned potential benefits will be realized. The observation that an orally bioavailable small-molecule inhibitor (BMS-488043), and a CD4-IgG fusion protein (PRO542) exhibit antiviral efficacy in the majority of infected patients studied is encouraging and further establishes gp120 as a potentially exploitable antiviral target. The clinical efficacy of the CD4-binding mAb TNX-355 is also encouraging. However, the promising antiviral effects of these drugs were observed in relatively short-term trials thus far. Due to the diversity of the HIV-1 Envelope, which reduces the anti-HIV spectrum and increases the occurrence of resistance to entry inhibitors, more comprehensive clinical studies will be needed to assess the impact on strain coverage and resistance development in patients when compounds in this class are co-administered with anti-HIV agents that operate through distinct mechanisms. Moreover, similar to the conformational basis of normal Envelope functions in the viral entry process, mechanisms of inhibition and resistance development to entry inhibitors are primarily dependent on Envelope conformational changes. Therefore, the effect of neutralizing antibodies present in patients, which also target the viral Envelope protein, on the development of resistance to entry inhibitors will require further study. Finally, approaches aimed at blocking HIV-1 entry via targeting of the cellular CD4 receptor must demonstrate that antiviral activity will not come at the expense of aberrant immune responses or side effects arising from the blockade of a normal host function. However, the encouraging safety data from early clinical studies with TNX-355 suggest that inhibition of CD4 binding may be achieved without inducing such responses. Therefore, it may be premature to exclude CD4 as a viable target for antiretroviral therapy.

Thus, the HIV-1 entry process offers multiple antiviral targets and provides emerging opportunities for drug development. This chapter has focused on selective gp120/CD4 interaction inhibitors that are in an advanced stage of study. Additional studies will be needed to ascertain whether the apparently potent inhibitors described in Pfizer's recent patent application or Chiron's efforts to identify epitope stabilizing compounds might lead to new candidates for clinical studies. In summary, preclinical and initial clinical results suggest that inhibitors preventing the interaction of gp120 with CD4 will provide exciting new treatment options and preventive measures for HIV-infected patients. Moreover, these inhibitors will serve as tools for studying HIV-1 entry events and for enhancing the understanding of the HIV-1 Envelope structure and function.

References

1 Kwong PD, Wyatt R, Robinson J, Sweet RW, Sodroski J, Hendrickson WA (1998) Structure of an HIV gp120 envelope glycoprotein in complex with the CD4 receptor and a neutralizing human antibody. *Nature* 393: 648–659

2 Wyatt R, Sodroski J (1998) The HIV-1 envelope glycoproteins: fusogens, antigens, and immunogens. *Science* 280: 1884–1888

3 Chen B, Vogan EM, Gong H, Skehel JJ, Wiley DC, Harrison SC (2005) Structure of an unliganded simian immunodeficiency virus gp120 core. *Nature* 433: 834–841

4 Hanna G, Lalezari J, Hellinger J, Wohl D, Masterson T, Fiske W, Kadow J, Lin P, Giordano M, Colonno R, Grasela D (2004) Antiviral activity, safety, and tolerability of a novel, oral small-molecule HIV-1 attachment inhibitor, BMS-488043, in HIV-1-infected subjects. In: 11th Conference on Retroviruses and Opportunistic Infections, San Francisco, abstract no. 141

5 Jacobson JM, Israel RJ, Lowy I, Ostrow NA, Vassilatos LS, Barish M., Tran DN, Sullivan BM, Ketas TJ, O'Neill TJ, Nagashima KA, Huang W, Petropoulos CJ, Moore JP, Maddon PJ, Olson WC (2004) Treatment of advanced human immunodeficiency virus type 1 disease with the viral entry inhibitor PRO 542. *Antimicrob Agents Chemother* 48: 423–429

6 Lin P-F, Blair W, Wang T, Spicer T, Guo Q, Zhou N, Gong Y-F, Wang H-GH, Rose R, Yamanaka G et al. (2003) A small molecule HIV-1 inhibitor that targets the HIV-1 envelope and inhibits CD4 receptor binding. *Proc Natl Acad Sci USA* 100: 11013–11018

7 Wang T, Zhang Z, Wallace OB, Deshpande M, Fang H, Yang Z, Zadjura LM, Tweedie DL, Huang S, Zhao F et al. (2003) Discovery of 4-Benzoyl-1-{(4-methoxy-1H-pyrrolo[2,3-b]pyridin-3-yl)oxoacetyl}-2- (R)-methylpiperazine (BMS-378806): A novel HIV-1 attachment inhibitor that interferes with CD4-gp120 interactions. *J Med Chem* 46: 4236–4239
8 Guo Q, Ho HT, Dicker I, Fan L, Zhou N, Friborg J, Wang T, McAuliffe BV, Wang, HG, Rose RE et al. (2003) Biochemical and genetic characterizations of a novel human immunodeficiency virus type 1 inhibitor that blocks gp120-CD4 interactions. *J Virol* 77: 10528–10536
9 Ho HT, Fan L, Nowicka-Sans B, McAuliffe B, Li CB, Yamanaka G, Zhou N, Fang H, Dicker I, Dalterio R et al. (2006) Envelope conformational changes induced by human immunodeficiency virus type 1 attachment inhibitors prevent CD4 binding and downstream entry events. *J Virol* 80: 4017–4025
10 Si Z, Madani N, Cox JM, Chruma JJ, Klein JC, Schon A, Phan N, Wang L, Biorn AC, Cocklin S et al. (2004) Small-molecule inhibitors of HIV-1 entry block receptor-induced conformational changes in the viral envelope glycoproteins. *Proc Natl Acad Sci USA* 101: 5036–5041.
11 Madani N, Perdigoto AL, Srinivasan K, Cox JM, Chruma JJ, LaLonde J, Head M, Smith AB III, Sodroski JG (2004) Localized changes in the gp120 envelope glycoprotein confer resistance to human immunodeficiency virus entry inhibitors BMS-806 and #155. *J Virol* 78: 3742–3752
12 Zhao Q, Ma L, Jiang S, Lu H, Liu S, He Y, Strick N, Neamati N, Debnath AK (2005) Identification of N-phenyl-N'-(2,2,6,6-tetramethyl-piperidin-4-yl)-oxalamides as a new class of HIV-1 entry inhibitors that prevent gp120 binding to CD4. *Virology* 339: 213–225
13 Middleton DS, Mowbray CE, Stephenson PT, Williams DH (2005) Preparation of piperazine derivatives for use in pharmaceutical compositions for the treatment of HIV infection. PCT Int Appl WO 2005016344
14 Fenwick DR, Middleton DS, Stephenson PT, Tran TD, Williams DH (2005) Preparation of piperazine and piperidine derivatives as anti-HIV-agents. PCT Int Appl WO 2005121094
15 Srivastava IK, Sharma V, Barnett SW, Ulmer J (2005) Env polypeptide complexed with CD4 mimetics to induce production of neutralizing antibodies against AIDS and disorders related to HIV infection. PCT Int Appl WO 2005121175
16 Wang J, Le N, Heredia A, Song H, Redfield R, Wang LX (2005) Modification and structure-activity relationship of a small molecule HIV-1 inhibitor targeting the viral envelope glycoprotein gp120. *Org Biomol Chem* 3: 1781–1786
17 Trkola A, Pomales A, Yuan H, Korber B, Maddon P, Allaway G, Katinger H, Barbas C 3rd, Burton D, Ho D (1995) Cross-clade neutralization of primary isolates of human immunodeficiency virus type 1 by human monoclonal antibodies and tetrameric CD4-IgG. *J Virol* 69: 6609–6617
18 Jacobson JM, Lowy I, Fletcher CV, O'Neill TJ, Tran DN, Ketas TJ, Trkola A, Klotman ME, Maddon PJ, Olson WC, Israel RJ (2000) Single-dose safety, pharmacology, and antiviral activity of the human immunodeficiency virus (HIV) type 1 entry inhibitor PRO 542 in HIV-infected adults. *J Infect Dis* 182: 326–329
19 Dey B, Del Castillo CS, Berger EA (2003) Neutralization of human immunodeficiency virus type 1 by sCD4-17b, a single-chain chimeric protein, based on sequential interaction of gp120 with CD4 and coreceptor. *J Virol* 77: 2859–2865
20 Vita C, Drakopoulou E, Vizzavona J, Rochette S, Martin L, Menez A, Roumestand C, Yang YS, Ylisastigui L, Benjouad A, Gluckman JC (1999) Rational engineering of a miniprotein that reproduces the core of the CD4 site interacting with HIV-1 envelope glycoprotein. *Proc Natl Acad Sci USA* 96: 13091–13096
21 Martin L, Stricher F, Misse D, Sironi F, Pugniere M, Barthe P, Prado-Gotor R, Freulon I, Magne X, Roumestand C et al. (2003) Rational design of a CD4 mimic that inhibits HIV-1 entry and exposes cryptic neutralization epitopes. *Nat Biotechnol* 21: 71–76
22 Li H, Song H, Heredia A, Le N, Redfield R, Lewis George K, Wang LX (2004) Synthetic bivalent CD4-mimetic miniproteins show enhanced anti-HIV activity over the monovalent miniprotein. *Bioconjug Chem* 15: 783–789
23 Samanen J (1998) Preparation of modified peptides and cyclopeptides as CD4 mimetic ligands for inhibiting HIV. PCT Int Appl WO 9826660 A1
24 Li C, Dowd CS, Zhang W, Chaiken IM (2001) Phage randomization in a charybdotoxin scaffold leads to CD4-mimetic recognition motifs that bind HIV-1 envelope through non-aromatic sequences. *J Pept Res* 57: 507–518
25 Binley JM, Wrin T, Korber B, Zwick MB, Wang M, Chappey C, Stiegler G, Kunert R, Zolla-Pazner S, Katinger H et al. (2004) Comprehensive cross-clade neutralization analysis of a panel

of anti-human immunodeficiency virus type 1 monoclonal antibodies. *J Virol* 78: 13232–13252
26 Saphire EO, Parren PW, Pantophlet R, Zwick MB, Morris GM, Rudd PM, Dwek RA, Stanfield RL, Burton DR, Wilson IA (2001) Crystal structure of a neutralizing human IgG against HIV-1: a template for vaccine design. *Science* 293: 1155–1159
27 Vermeire K, Schols D (2005) Cyclotriazadisulfonamides: promising new CD4-targeted anti-HIV drugs. *J Antimicrob Chemother* 56: 270–272
28 Yang QE, Stephen AG, Adelsberger JW, Roberts PE, Zhu W, Currens MJ, Feng Y, Crise BJ, Gorelick RJ, Rein AR et al. (2005) Discovery of small-molecule human immunodeficiency virus type 1 entry inhibitors that target the gp120-binding domain of CD4. *J Virol* 79: 6122–6133
29 Rishi V, Potter T, Laudeman J, Reinhart R, Silvers T, Selby M, Stevenson T, Krosky P, Stephen AG, Acharya A et al. (2005) A high-throughput fluorescence-anisotropy screen that identifies small molecule inhibitors of the DNA binding of B-ZIP transcription factors. *Anal Biochem* 340: 259–271
30 Moore JP, Sattentau QJ, Klasse PJ, Burkly LC (1992) A monoclonal antibody to CD4 domain 2 blocks soluble CD4-induced conformational changes in the envelope glycoproteins of human immunodeficiency virus type 1 (HIV-1) and HIV-1 infection of $CD4^+$ cells. *J Virol* 66: 4784–4793
31 Reimann KA, Lin W, Bixler S, Browning B, Ehrenfels BN, Lucci J, Miakowski K, Olson D, Parish TH, Rosa MD et al. (1997) A humanized form of a CD4-specific monoclonal antibody exhibits decreased antigenicity and prolonged plasma half-life in rhesus monkeys while retaining its unique biological and antiviral properties. *AIDS Res Hum Retroviruses* 13: 933–943
32 Kuritzkes DR, Jacobson J, Powderly WG, Godofsky E, DeJesus E, Haas F, Reimann KA, Larson JL, Yarbough PO, Curt V, Shanahan WR Jr (2004) Antiretroviral activity of the anti-CD4 monoclonal antibody TNX-355 in patients infected with HIV type 1. *J Infect Dis* 189: 286–291
33 Nagashima KA, Thompson DA, Rosenfield SI, Maddon PJ, Dragic T, Olson WC (2001) Human immunodeficiency virus type 1 entry inhibitors PRO 542 and T-20 are potently synergistic in blocking virus-cell and cell-cell fusion. *J Infect Dis* 183: 1121–1125
34 Tremblay CL, Kollmann C, Giguel F, Chou TC, Hirsch MS (2000) Strong *in vitro* synergy between the fusion inhibitor T-20 and the CXCR4 blocker AMD-3100. *J Acquir Immune Defic Syndr* 25: 99–102

Entry Inhibitors in HIV Therapy
Edited by Jacqueline D. Reeves and Cynthia A. Derdeyn

Inhibitors that target gp120 interactions with coreceptor

Julie M. Strizki[1] and Donald E. Mosier[2]

[1] *Department of Virology, Schering-Plough Research Institute, 2015 Galloping Hill Road, K15, E405C/4945, Kenilworth, NJ 07033, USA*
[2] *Department of Immunology, The Scripps Research Institute, 10550 North Torrey Pines Road, La Jolla, California 92037, USA*

Virus entry, cell tropism, and coreceptor switching

Until 1996, when HIV-1 infection was discovered to involve chemokine receptors (reviewed in [1, 2]), virus isolates were characterized by biological assays on human cell lines as syncytium inducing (SI) or non-syncytium inducing (NSI) [3]. NSI viruses predominate early after transmission, are capable of growth in primary macrophages (macrophage tropism), and fail to grow in most established human T cell lines. SI viruses, by contrast, are found in only a subset of patients after chronic infection, are usually not macrophage tropic, and infect established T cell lines [4]. NSI viruses were found to use CCR5 as the coreceptor for virus entry into target cells, and SI viruses to use CXCR4 for entry, either alone or, more commonly, in addition to CCR5. Since CCR5 and CXCR4 are differentially expressed on the surface of naïve and memory T cells, dendritic cells, and macrophages, coreceptor preference largely explains cell tropism. The importance of CCR5 in viral transmission and as a target for therapeutic intervention was underscored by the discovery of a 32-base pair deletion in the human CCR5 coding region (Δ32 mutation) which, when homozygous, prevents CCR5 surface expression and HIV-1 infection [5, 6]. Individuals with the CCR5 Δ32/Δ32 genotype appear to suffer no clinically obvious detriment, although recent data suggest that they may be more susceptible to West Nile virus encephalitis [7]. The cell tropism of CCR5-using viruses (R5 isolates) probably helps explain their selective transmission [8], although it is still difficult to understand why R5 isolates predominate after parenteral exposure [9]. High virus loads during unrecognized primary infection may be one factor that contributes to higher transmission rates for R5 isolates, and dendritic cells appear to be more susceptible to R5 virus infection [10, 11]. As HIV-1 infection progresses and CD4 T cell numbers decline, the fraction of patients with HIV-1 isolates capable of using CXCR4 (R5X4 or X4) increases to nearly 50% at end-stage disease [12]. Because many experiments fail to distinguish between R5X4 and X4 viruses, we will use (R5)X4 to indicate HIV-1 isolates that can infect via CXCR4. This switch in coreceptor use has major implications for the use of CCR5 inhibitors, since they may select for HIV-1 minor populations capable of CXCR4 use, and such viruses are associated with a more rapid disease progression [13]. Indeed, short-term monotherapy with maraviroc (see Tab. 1) did select for R5X4 and X4 variants in 2/62 patients [14], although these viruses appear to have emerged from a preexisting reservoir and not from *de novo* mutations. Nonetheless, these results underscore the risk of selectively inhibiting CCR5 and highlight the need to use these agents in combination with other potent antivirals. The mutations in the variable loops of envelope that drive coreceptor switching often come at the cost of viral fitness [15, 16], which may explain the long delay until the appearance of R5X4 or X4 viruses in patients, and also why 60/62 patients treated with maraviroc did not show emergence of R5X4 or X4 variants [14]. Envelope mutations occur throughout the course of infection [17], and appear to increase resistance to coreceptor inhibitors before [18] and after [19] coreceptor switching.

CCR5 and CXCR4 expression and sensitivity to inhibition

Understanding the biology of the chemokine receptors is important for appreciating the activity of inhibitors targeting these receptors. CCR5 and CXCR4 are integral membrane proteins with seven membrane-spanning domains and four extracellular domains, the N terminus, and extracellular loops 1, 2, and 3 (ECL1, 2, and 3). Their normal function is to bind chemokines and signal cell movement (chemotaxis) via coupled G proteins, and chemokine binding normally leads to receptor internalization mediated by binding of the serine phosphorylated intracellular C terminus to β-arrestins [20]. While CCR5 shows the typical promiscuity of chemokine receptors and binds MCP-2 (CCL8), MCP-5 (CCL12), MIP 1β (CCL4), MIP-1α (CCL3), LD-78β (CCL3L1), RANTES (CCL5) and C10 (CCL6), CXCR4 binds only SDF-1 α/β (CXCL12) [21]. Mutations that block CCR5 expression are well tolerated in both humans and mice, but deletion of CXCR4 expression is lethal in mice [22]. Native chemokine binding to CCR5 or CXCR4 is sufficient to inhibit HIV-1 infection [23, 24], although the potency of inhibition is substantially less than that achieved by candidate therapeutics targeting the coreceptors. Inhibition of HIV-1 infection by chemokines is associated with both receptor blockade and receptor internalization (Fig. 1) [24–26]. Receptor internalization is a prominent feature of most N-terminal modifications of chemokines, e.g., AOP-RANTES [25] or PSC-RANTES [27], and correlates with their *in vitro* potency. However, some modified chemokines retain antiviral activity in the absence of receptor internalization [28]. Different classes of CCR5 and CXCR4 inhibitors bind to the coreceptors in distinct manners. Most small-molecule CCR5 inhibitors appear to bind to a hydrophobic pocket in the upper transmembrane spanning region that is flanked by several membrane-spanning

Table 1. Coreceptor inhibitors

Compound	Manufacturer	Identity	Target	Phase	Reference
Met-RANTES	N/A	N-terminal methionine added	CCR5	Preclinical	[51]
RANTES (9-68)	N/A	Truncated RANTES	CCR5	Preclinical	[78]
RANTES (3-68)	N/A	CD26-processed RANTES	CCR5	Preclinical	[79]
AOP-RANTES	N/A	Aminooxypentane-RANTES	CCR5	Preclinical	[80]
NNY-RANTES	N/A	*N*-nonanoyl-RANTES	CCR5	Preclinical	[57]
PSC-RANTES	N/A	*N*-nonanoyl, thioproline, cyclohexylglycine-RANTES	CCR5	Preclinical	[27]
LD78β	N/A	Isoform of MIP-1α	CCR5	Preclinical	[81]
Met-SDF1	N/A	N-terminal methionine added	CXCR4	Preclinical	[82]
P1/P2-RANTES	N/A	Phage display selection of mutant library	CCR5	Preclinical	[83]
RCP189	N/A	Modified chemokine	CCR5	Preclinical	[28]
RCP169	N/A	Modified chemokine	CXCR4	Preclinical	[28]
ALX40-4C	N/A	9 D-amino acid peptide	CXCR4	I	[58, 59]
T22	N/A	18 amino acid peptide	CXCR4	Preclinical	[84]
T140	N/A	14 amino acid peptide	CXCR4	Preclinical	[60]
CCL5 + FvCD4	N/A	Fusion protein of RANTES-anti-CD4	CCR5 + CD4	Preclinical	[85]
PRO-140	Progenics	Monoclonal antibody	CCR5	I	[38]
CCR5mAb004	Human Genome Sciences	Monoclonal antibody	CCR5	I	[86]
AMD3100	AnorMed	Bicyclam	CXCR4	Discontinued	[72]
AMD3465	AnorMed	Monomacrocyclic	CXCR4	Preclinical	[87]
AMD3451	AnorMed	*N*-pyridinylmethyl cyclam	CCR5 & CXCR4	Preclinical	[88]
AMD887	AnorMed	Small molecule	CCR5	Preclinical	[67]

(Continued on next page)

Table 1. (Continued)

Compound	Manufacturer	Identity	Target	Phase	Reference
AMD070	AnorMed	Small molecule	CXCR4	I/II	[89]
"9 g"	AstraZeneca	Phenylacetamide	CCR5	Preclinical	[90]
GW873140 (Aplaviroc)	Glaxo SmithKline	Spirodiketopiperazine-derivative	CCR5	Discontinued	[91]
KRH-2731	Kureha	Arginine derivative	CXCR4	Preclinical	[68]
KRH-3955	Kureha		CXCR4	Preclinical	[69]
KRH-3140	Kureha		CXCR4	Preclinical	[69]
UK-427,857 (Maraviroc)	Pfizer	Benzimidazole, 4,4-fluorcyclohexanecarboxamide	CCR5	II/III	[92]
Sch-C	Schering-Plough	Oximino-piperdicino-piperidine amide	CCR5	Discontinued	[93]
Sch-D (Vicriviroc)	Schering-Plough	Piperazino-piperdine	CCR5	II	[94]
INCB9471	Incyte		CCR5	Preclinical	See [a]
Cmpd 167	Merck		CCR5	Preclinical	[77]
TAK-779	Takeda	Quaternary ammonium anilide	CCR5	Discontinued	[64]
TAK-220	Takeda	Piperdine-4-carboxamide	CCR5	Preclinical	[95]
TAK-652	Takeda	Carboxamide monomethanesulfonate	CCR5	Preclinical	[96]

[a] http://www.incyte.com/drugs_product_pipeline.html

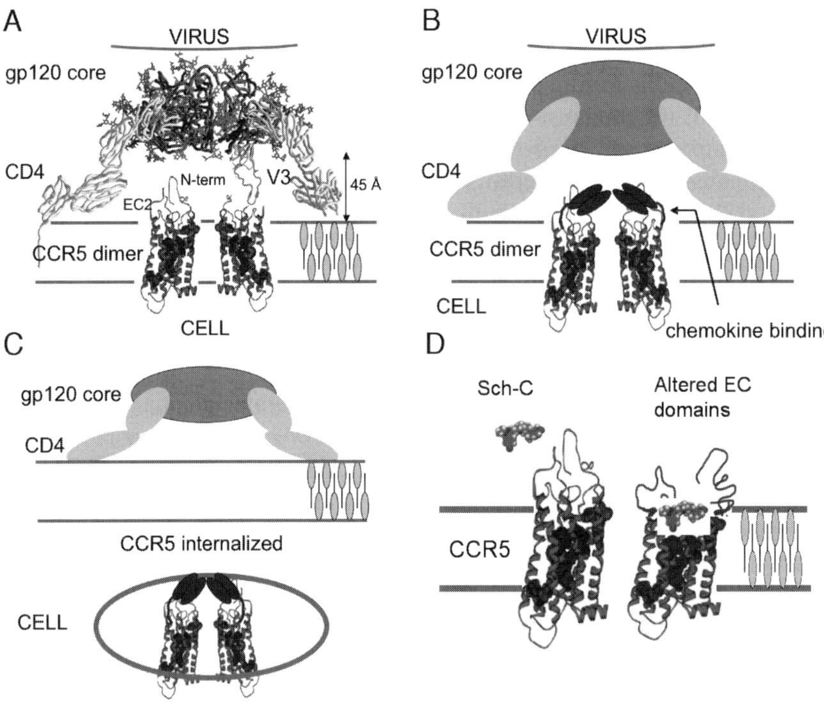

Figure 1. Mechanisms of CCR5 inhibition. (A) HIV-1 envelope trimer first binds CD4, then undergoes a conformation change to permit CCR5 (or CXCR4) binding. Molecules are drawn to scale from structural studies [97–99]. The distance from the HIV gp120 core structure to the cell membrane is 45 Å, and the V3 loop extends about 30 Å towards the cell, permitting extensive binding to extracellular domains of CCR5. (B) Native or modified chemokines or anti-CCR5 antibodies bind to the extracellular domains of CCR5 and prevent access to HIV-1 gp120. (C) In addition to steric hindrance, binding of native and particularly modified chemokines induces CCR5 internalization and sequestration. (D) Small molecule inhibitors appear to bind to hydrophobic pockets in the CCR5 membrane spanning domains and alter the conformations of the extracellular domains.

domains and the base of ECL2 [29–31]. This binding presumably results in a conformational change in the extracellular domains of CCR5. A similar mechanism of action has been proposed for AMD3100 binding to CXCR4 [32, 33]. Modified chemokines (e.g., RANTES) are thought to bind via two sites [34]; the N-terminal domain (residues 1–10) appears to insert adjacent to the transmembrane domains, while the conserved chemokine core structure binds ECL1-3. The CCR5 N-terminal and ECL2 domains are most critical for HIV-1 infection [35], while the N-terminal region of CXCR4 is less critical for (R5)X4 virus infection [36]. Anti-CCR5 antibodies that recognize the ECL2 domain are the most potent at inhibiting HIV-1 infection [37, 38]. While all the coreceptor inhibitors share the ability to interfere with the binding of CD4-triggered gp120 to CCR5 or CXCR4, they may accomplish this goal by different mechanisms. Small-molecule antagonists likely act by disrupting the conformation of extracellular domains, and modified chemokines and antibodies block access to the coreceptor by steric inhibition or by receptor internalization (modified chemokines). Gene therapy strategies that reduce coreceptor expression have also been explored [39–41].

Binding of the gp120 V3 loop and the conserved bridging sheet to CCR5 can vary extensively between different HIV-1 envelopes [42, 43]. Likewise, the conformation of individual CCR5 molecules expressed on a single cell may differ [44], and the amount of CCR5 expressed is controlled by polymorphisms in the CCR5 promoter region, the coding region, and in genes encoding chemokine ligands. This variability in HIV-1 envelope binding to CCR5 likely accounts for the large variation in 50% inhibitory concentrations (IC_{50}) observed when multiple HIV-1 isolates are tested against a single CCR5 inhibitor [45–47]. Similar arguments apply to the binding of HIV-1 envelope to CXCR4 [19, 48].

Chemokine-based and peptide coreceptor inhibitors

Following the recognition that chemokines could inhibit HIV-1 infection of T lymphocytes [23], but were poor at preventing infection of macrophages [49], more potent modification of chemokines, especially RANTES (CCL5), were developed (Tab. 1). Modification of the N terminus of RANTES, either by truncation or by chemical modification, resulted in compounds with greater inhibitory activity on all target cells [25, 50]. The improvement in the activity of AOP-RANTES was attributed to its greater affinity for CCR5 as well as its improved ability to internalize CCR5 [25, 51]. Similar effects were seen when the N terminus of MIP-1α or LD78β was modified with AOP [52]. However, subsequent N-terminal modifications of RANTES (e.g., NNY- or PSC-RANTES; see Table 1) increased the potency of HIV-1 inhibition *in vitro* to the picomolar range without changing CCR5 binding affinity [27]. Instead, the potency of inhibition correlated with the extent and duration of CCR5 internalization (Fig. 1) [27, 53, 54]. PSC-RANTES is advancing as a candidate microbicide, and is capable of preventing vaginal transmission of the R5 SHIV162P3 in rhesus macaques [55]. The agonist activity of these compounds raises some concerns about side effects, but modified chemokines without agonist activity are also being developed [28, 56]. The ability of this class of inhibitors to sequester CCR5 within the target cell may provide a barrier to the development of resistance. To date, no resistant variants to PSC-RANTES have been selected *in vitro*, although two independent R5X4 coreceptor switch variants were selected during treatment with the less potent NNY-RANTES in hu-PBL-SCID mouse experiments [57].

Several peptide inhibitors that target CXCR4 have been identified (Tab. 1). ALX40-4C was initially developed to inhibit the Tat-TAR interaction, but was subsequently found to bind ECL2 of CXCR4 and block X4 HIV-1 infection [58, 59]. This compound was well tolerated and safe in Phase I clinical trials, but failed to reduce viral load even in those patients harboring (R5)X4 virus [58]. T22 and the more potent T140 derivative bind to CXCR4 and inhibit SDF1 binding as well as X4 HIV-1 infection [60]. Development of more potent and biostable derivatives of T140 continues [61].

Antibodies targeting coreceptors

Two anti-CCR5 antibodies are in Phase I clinical trials, Pro 140 and CCR5mAb004 (Tab. 1). The results of the Progenics trial showed a dose-dependent binding of PRO 140 to CCR5-expressing cells, with the highest PRO 140 concentration tested coating CCR5 cells for at least 60 days following a single infusion [62]. The CCR5mAb004 is a fully humanized immunoglobulin generated in the Abgenix Xenomouse model. Phase I results are not yet available. Because of the relatively large size of intact immunoglobulins compared to the CD4-envelope complex, antibodies must bind to CCR5 prior to virus exposure to be effective [63].

Small-molecule coreceptor inhibitors

Following the discovery that the chemokine receptors, CCR5 and CXCR4 were essential for HIV infection and that the ligands for these receptors had antiviral activity, many groups began to search for small-molecule receptor antagonists as new agents for antiviral therapy. The first small-molecule CCR5 antagonist to be described was TAK-779 [64]. This compound had potent antiviral activity *in vitro* against R5-tropic strains, but lacked adequate bioavailability and was not further developed. Other groups also identified potent small-molecule inhibitors of CCR5 (listed in Table 1), including: TAK-220, TAK-652, Cpmd 167, SCH-C, vicriviroc (SCH 417690, SCH-D), maraviroc (UK-427,857) and aplaviroc (GW-873140). Interestingly, many of these compounds are structurally distinct, although their binding sites map to a similar pocket formed at the base of the extracellular loops of CCR5 [29, 31]. Binding of these inhibitors to CCR5 is believed to cause a conformational change in the receptor such that CD4-bound gp120 can no longer engage CCR5 efficiently (Fig. 1). Comparative binding and mutagenesis studies suggest that individual inhibitors induce slightly different receptor conformations by interacting with different residues within CCR5. Although all these small molecules are functional antagonists of CCR5, differential binding properties may have implications for reduced cross-resistance between compounds from different structural classes.

Resistance to small-molecule inhibitors has been achieved by continuous passage of virus in the presence of compound *in vitro*. Unlike the viral protease and reverse transcriptase enzymes, which have structurally conserved active sites, the viral gp120 protein is highly variable and different conformations appear to be capable of engaging the coreceptor. Therefore, it is not surprising that individual viral isolates accumulate different mutational patterns, both within the V3 loop and elsewhere, that are associated with reduced susceptibility to CCR5 antagonists [14, 38, 65]. For these reasons, resistance is measured phenotypically as no genotypic algorithm is currently available to predict susceptibility to this class of compounds. Additional data from clinical trials will be needed to better understand the genotypic and phenotypic correlates of resistance to these inhibitors.

In addition to CCR5 inhibitors, several groups have identified small-molecule antagonists of the CXCR4 receptor. AMD 3100, a bicyclam, was the first small-molecule CXCR4 antagonist to demonstrate antiviral activity *in vitro* [66]. However, this molecule is not orally bioavailable and is no longer being developed for HIV-1 therapy. A second-generation molecule with greater potency and bioavailability, AMD 070, was subsequently developed by AnorMed and is currently in clinical trials [67]. Other inhibitors of CXCR4 that have been described are KRH 1636 [68], KRH 3955 and KRH 3140 by Kureha [69]. The latter two compounds can be dosed orally and have demonstrated efficacy in protecting hu-PBL-SCID mice from infection [69].

Clinical status of coreceptor antagonists

Although numerous small molecule coreceptor inhibitors have been described, only a few have entered clinical trials to date. The CCR5 antagonists SCH-C, vicriviroc, maraviroc and aplavaviroc all demonstrated potent antiviral activity in short-term monotherapy trials, with the latter three compounds all capable of achieving mean viral load reductions of >1.6 logs following 10–14 days of dosing (Fig. 2). Interestingly, in each of these monotherapy studies, a pro-

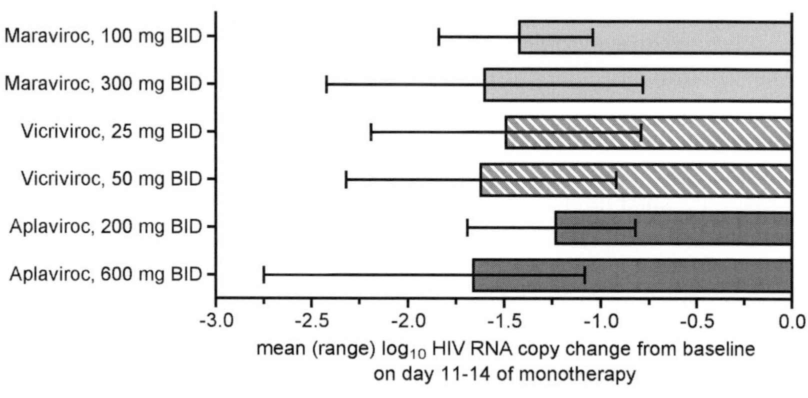

Figure 2. Mean reduction in HIV-1 plasma RNA copy number during monotherapy trials of CCR5 inhibitors. Data are derived from references [71, 75, 100].

longed period of viral suppression was observed during the washout period of the trial. This effect is believed to be due in part to the high affinity of the compounds and the slow off-rate from the CCR5 receptor. Although all the CCR5 inhibitors have demonstrated activity, development of two of these compounds, SCH-C and aplaviroc, has been halted due to adverse events. In the case of SCH-C, changes in the cardiac QTc interval were noted in the higher dose groups and several cases of acute liver toxicity were seen in the aplaviroc trial [70]. However, neither of these effects is believed to be based on the actual mechanism as they have not been reported for other CCR5 inhibitors.

Currently, both maraviroc and vicriviroc are being evaluated for safety, efficacy and durability in Phase II and Phase III trials in treatment-experienced subjects in the presence of an optimized background regimen. In addition, smaller studies have been initiated in treatment-naïve patients to evaluate efficacy in this population in combination with two nucleoside reverse transcriptase inhibitors. In the vicriviroc study, treatment-naïve subjects initially responded well to the treatment; however, following 4–5 months of therapy, a dose-related increase in the incidence of viral rebound was observed in the vicriviroc arms and the study was stopped [71]. In a similarly designed trial of maraviroc, the low-dose arm of the study was stopped due to inferior response, although the study is continuing at the higher dose. Future studies of CCR5 inhibitors in treatment-naïve patients may require more careful selection of dose and background regimen to ensure adequate viral suppression.

The clinical development of CXCR4 antagonists has proven more challenging relative to the CCR5 inhibitors. The first small-molecule CXCR4 inhibitor, AMD3100, was tested in a proof-of-concept study in a cohort of patients with single or dual/mixed tropic isolates [72]. Although the study goal of 1 log reduction in viral load was not met and the trial subsequently halted, follow up studies revealed a reduction in X4-tropic variants in the plasma of treated patients, providing validation of CXCR4 as a target. Interestingly, patients receiving AMD3100 experienced significant increases in their leukocyte counts. This mechanism-based effect is believed to result from inhibition of the interaction of SDF-1 on the vascular endothelium and CXCR4 on leukocytes, causing mobilization of the cells. An orally bioavailable backup compound, AMD070, is now in Phase II studies [67].

Optimal applications of coreceptor inhibitors

Several questions remain regarding how, when and in which patient populations coreceptor inhibitors should be used for antiretroviral therapy. Since R5-tropic viruses predominate in treatment-naïve individuals early in the course of infection, it may be appropriate to include a CCR5 inhibitor as part of first line therapy. However, the safety and durability of response to antiretroviral regimens containing CCR5 inhibitors in naïve patients is still under investigation. Although early clinical trials excluded patients with X4-tropic viruses, the currently available tropism assays [73, 74] can miss the presence of minor populations of X4-tropic viruses, which led to a suboptimal response in some patients in a monotherapy trial [14, 75]. In addition, combination therapy with CCR5 antagonists and AZT/3TC led to premature viral breakthrough in the lower dose arms following several months of therapy ([71] and 24 January 2006 Pfizer press release). It will thus be important to establish the safety, efficacy and optimal dosing and background regimen for coreceptor inhibitors in this patient population as these agents continue to be developed.

The use of CCR5 and CXCR4 antagonists in the more advanced, treatment-experienced population will provide more treatment options to patients, but also poses some unique challenges. Because these compounds have a novel mechanism of action, pre-existing drug resistance within the patient population is expected to be low. However, approximately 50% of treatment-experienced patients have detectable (R5)X4 viruses in their plasma. Currently, it is not known whether CCR5 or CXCR4 inhibitors in combination with other antivirals will prove safe and effective in these patients. Additional studies will be necessary to better define the relationship between viral tropism and treatment outcome in experienced patients. Consequently, viral tropism testing will likely be necessary prior to and during therapy to monitor changes in coreceptor use until the safety and efficacy of treatment regimens containing coreceptor inhibitors are better established in the clinic.

Ideally, the use of combinations of CCR5 and CXCR4 inhibitors would avoid the problem of selecting for (R5)X4 variants. However, the current regulatory practices would require each of the component agents to demonstrate safety and efficacy independently, effectively delaying a very attractive option. Molecules such as AMD3451 (see Tab. 1) that target both CCR5 and CXCR4 offer a solution to this obstacle, and additional candidates with these properties would be a welcome development.

If current clinical trials demonstrate safety of candidate CCR5 inhibitors, they may be very useful in microbicide formulations and in pre- and post-exposure prophylaxis applications. The predominance of R5 viruses in primary transmission makes CCR5 inhibition a logical strategy to prevent HIV infection. In fact, animal studies with the Merck CCR5 antagonist Cpmd 167 have demonstrated protection from vaginal challenge in monkeys treated both topically and orally with the compound [76, 77]. Although these studies are encouraging, additional carefully controlled clinical trials will be needed to demonstrate the effectiveness and safety of these agents for prophylaxis of infection.

In summary, small molecule inhibitors of the HIV coreceptors CCR5 and CXCR4 have demonstrated potent *in vitro* and *in vivo* activity and several compounds are advancing in clinical trials. Although these compounds will likely prove useful as part of an antiretroviral regimen, their optimal use during the course of HIV treatment has yet to be clearly defined.

References

1 Moore JP (1997) Coreceptors: implications for HIV pathogenesis and therapy. *Science* 276: 51–52

2 Berger EA, Murphy PM, Farber JM (1999) Chemokine receptors as HIV-1 coreceptors: roles in viral entry, tropism, and disease. *Annu Rev Immunol* 17: 657–700

3 Tersmette M, Lange JM, de Goede RE, de Wolf F, Eeftink-Schattenkerk JK, Schellekens PT, Coutinho RA, Huisman JG, Goudsmit J, Miedema F (1989) Association between biological properties of human immunodeficiency virus variants and risk for AIDS and AIDS mortality. *Lancet* 1: 983–985

4 de Roda Husman AM, Schuitemaker H (1998) Chemokine receptors and the clinical course of HIV-1 infection. *Trends Microbiol* 6: 244–249

5 Huang Y, Paxton WA, Wolinsky SM, Neumann AU, Zhang L, He T, Kang S, Ceradini D, Jin Z, Yazdanbakhsh K et al. (1996) The role of a mutant CCR5 allele in HIV-1 transmission and disease progression. *Nat Med* 2: 1240–1243

6 Dean M, Carrington M, Winkler C, Huttley GA, Smith MW, Allikmets R, Goedert JJ, Buchbinder SP, Vittinghoff E, Gomperts E et al. (1996) Genetic restriction of HIV-1 infection and progression to AIDS by a deletion allele of the CKR5 structural gene. Hemophilia Growth and Development Study, Multicenter AIDS Cohort Study, Multicenter Hemophilia Cohort Study, San Francisco City Cohort, ALIVE Study. *Science* 273: 1856–1862

7 Glass WG, McDermott DH, Lim JK, Lekhong S, Yu SF, Frank WA, Pape J, Cheshier RC, Murphy PM (2006) CCR5 deficiency increases risk of symptomatic West Nile virus infection. *J Exp Med* 203: 35–40

8 Margolis L, Shattock R (2006) Selective transmission of CCR5-utilizing HIV-1: the 'gatekeeper' problem resolved? *Nat Rev Microbiol* 4: 312–317

9 Wilkinson DA, Operskalski EA, Busch MP, Mosley JW, Koup RA (1998) A 32-bp deletion within the CCR5 locus protects against transmission of parenterally acquired human immunodeficiency virus but does not affect progression to AIDS-defining illness. *J Infect Dis* 178: 1163–1166

10 Kawamura T, Gulden FO, Sugaya M, McNamara DT, Borris DL, Lederman MM, Orenstein JM, Zimmerman PA, Blauvelt A (2003) R5 HIV productively infects Langerhans cells, and infection levels are regulated by compound CCR5 polymorphisms. *Proc Natl Acad Sci USA* 100: 8401–8406

11 Smed-Sorensen A, Lore K, Vasudevan J, Louder MK, Andersson J, Mascola JR, Spetz AL, Koup RA (2005) Differential susceptibility to human immunodeficiency virus type 1 infection of myeloid and plasmacytoid dendritic cells. *J Virol* 79: 8861–8869

12 Brumme ZL, Goodrich J, Mayer HB, Brumme CJ, Henrick BM, Wynhoven B, Asselin JJ, Cheung PK, Hogg RS, Montaner JS et al. (2005) Molecular and clinical epidemiology of CXCR4-using HIV-1 in a large population of antiretroviral-naive individuals. *J Infect Dis* 192: 466–474

13 Paxton WA, Kang S (1998) Chemokine receptor allelic polymorphisms: relationships to HIV resistance and disease progression. *Semin Immunol* 10: 187–194

14 Westby M, Lewis M, Whitcomb J, Youle M, Pozniak AL, James IT, Jenkins TM, Perros M, van der Ryst E (2006) Emergence of CXCR4-using human immunodeficiency virus type 1 (HIV-1) variants in a minority of HIV-1-infected patients following treatment with the CCR5 antagonist maraviroc is from a pretreatment CXCR4-using virus reservoir. *J Virol* 80: 4909–4920

15 Pastore C, Ramos A, Mosier DE (2004) Intrinsic obstacles to human immunodeficiency virus type 1 coreceptor switching. *J Virol* 78: 7565–7574

16 Pastore C, Nedellec R, Ramos A, Pontow S, Ratner L, Mosier DE (2006) Human immunodeficiency virus type 1 coreceptor switching: V1/V2 gain-of-fitness mutations compensate for V3 loss-of-fitness mutations. *J Virol* 80: 750–758

17 Shankarappa R, Margolick JB, Gange SJ, Rodrigo AG, Upchurch D, Farzadegan H, Gupta P, Rinaldo CR, Learn GH, He X et al. (1999) Consistent viral evolutionary changes associated with the progression of human immunodeficiency virus type 1 infection. *J Virol* 73: 10489–10502

18 Gorry PR, Sterjovski J, Churchill M, Witlox K, Gray L, Cunningham A, Wesselingh S (2004) The role of viral coreceptors and enhanced macrophage tropism in human immunodeficiency virus type 1 disease progression. *Sex Health* 1: 23–34

19 Stalmeijer EH, Van Rij RP, Boeser-Nunnink B, Visser JA, Naarding MA, Schols D, Schuitemaker

H (2004) *In vivo* evolution of X4 human immunodeficiency virus type 1 variants in the natural course of infection coincides with decreasing sensitivity to CXCR4 antagonists. *J Virol* 78: 2722–2728
20 Oppermann M (2004) Chemokine receptor CCR5: insights into structure, function, and regulation. *Cell Signal* 16: 1201–1210
21 Wells TN, Power CA, Shaw JP, Proudfoot AE (2006) Chemokine blockers – therapeutics in the making? *Trends Pharmacol Sci* 27: 41–47
22 Tachibana K, Hirota S, Iizasa H, Yoshida H, Kawabata K, Kataoka Y, Kitamura Y, Matsushima K, Yoshida N, Nishikawa S et al. (1998) The chemokine receptor CXCR4 is essential for vascularization of the gastrointestinal tract. *Nature* 393: 591–594
23 Cocchi F, DeVico AL, Garzino-Demo A, Arya SK, Gallo RC, Lusso P (1995) Identification of RANTES, MIP-1 alpha, and MIP-1 beta as the major HIV-suppressive factors produced by $CD8^+$ T cells *Science* 270: 1811–1815
24 Amara A, Gall SL, Schwartz O, Salamero J, Montes M, Loetscher P, Baggiolini M, Virelizier JL, Arenzana-Seisdedos F (1997) HIV coreceptor downregulation as antiviral principle: SDF-1alpha-dependent internalization of the chemokine receptor CXCR4 contributes to inhibition of HIV replication. *J Exp Med* 186: 139–146
25 Mack M, Luckow B, Nelson PJ, Cihak J, Simmons G, Clapham PR, Signoret N, Marsh M, Stangassinger M, Borlat F et al. (1998) Aminooxypentane-RANTES induces CCR5 internalization but inhibits recycling: a novel inhibitory mechanism of HIV infectivity. *J Exp Med* 187: 1215–1224
26 Pelchen-Matthews A, Signoret N, Klasse PJ, Fraile-Ramos A, Marsh M (1999) Chemokine receptor trafficking and viral replication. *Immunol Rev* 168: 33–49
27 Hartley O, Gaertner H, Wilken J, Thompson D, Fish R, Ramos A, Pastore C, Dufour B, Cerini F, Melotti A et al. (2004) Medicinal chemistry applied to a synthetic protein: development of highly potent HIV entry inhibitors. *Proc Natl Acad Sci USA* 101: 16460–16465
28 Kumar S, Choi WT, Dong CZ, Madani N, Tian S, Liu D, Wang Y, Pesavento J, Wang J, Fan X et al. (2006) SMM-chemokines: a class of unnatural synthetic molecules as chemical probes of chemokine receptor biology and leads for therapeutic development. *Chem Biol* 13: 69–79
29 Dragic T, Trkola A, Thompson DA, Cormier EG, Kajumo FA, Maxwell E, Lin SW, Ying W, Smith SO, Sakmar TP et al. (2000) A binding pocket for a small molecule inhibitor of HIV-1 entry within the transmembrane helices of CCR5. *Proc Natl Acad Sci USA* 97: 5639–5644
30 Maeda K, Das D, Ogata-Aoki H, Nakata H, Miyakawa T, Tojo Y, Norman R, Takaoka Y, Ding J, Arnold GF et al. (2006) Structural and molecular interactions of CCR5 inhibitors with CCR5. *J Biol Chem* 281: 12688–12698
31 Seibert C, Ying W, Gavrilov S, Tsamis F, Kuhmann SE, Palani A, Tagat JR, Clader JW, McCombie SW, Baroudy BM et al. (2006) Interaction of small molecule inhibitors of HIV-1 entry with CCR5. *Virology* 349: 41–54
32 Donzella GA, Schols D, Lin SW, Este JA, Nagashima KA, Maddon PJ, Allaway GP, Sakmar TP, Henson G, De Clercq E et al. (1998) AMD3100, a small molecule inhibitor of HIV-1 entry via the CXCR4 co- receptor. *Nat Med* 4: 72–77
33 Gerlach LO, Skerlj RT, Bridger GJ, Schwartz TW (2001) Molecular interactions of cyclam and bicyclam non-peptide antagonists with the CXCR4 chemokine receptor. *J Biol Chem* 276: 14153–14160
34 Blanpain C, Doranz BJ, Bondue A, Govaerts C, De Leener A, Vassart G, Doms RW, Proudfoot A, Parmentier M (2003) The core domain of chemokines binds CCR5 extracellular domains while their amino terminus interacts with the transmembrane helix bundle. *J Biol Chem* 278: 5179–5187
35 Rucker J, Doms RW (1998) Chemokine receptors as HIV coreceptors: implications and interactions. *AIDS Res Hum Retroviruses* 14 Suppl 3: S241–246
36 Chabot DJ, Zhang PF, Quinnan GV, Broder CC (1999) Mutagenesis of CXCR4 identifies important domains for human immunodeficiency virus type 1 X4 isolate envelope-mediated membrane fusion and virus entry and reveals cryptic coreceptor activity for R5 isolates. *J Virol* 73: 6598–6609
37 Olson WC, Rabut GE, Nagashima KA, Tran DN, Anselma DJ, Monard SP, Segal JP, Thompson DA, Kajumo F, Guo Y et al. (1999) Differential inhibition of human immunodeficiency virus type 1 fusion, gp120 binding, and CC-chemokine activity by monoclonal antibodies to CCR5. *J Virol* 73: 4145–4155
38 Trkola A, Ketas TJ, Nagashima KA, Zhao L, Cilliers T, Morris L, Moore JP, Maddon PJ, Olson

WC (2001) Potent, broad-spectrum inhibition of human immunodeficiency virus type 1 by the CCR5 monoclonal antibody PRO 140. *J Virol* 75: 579–588
39 Anderson J, Akkina R (2005) HIV-1 resistance conferred by siRNA cosuppression of CXCR4 and CCR5 coreceptors by a bispecific lentiviral vector. *AIDS Res Ther* 2: 1
40 Cordelier P, Kulkowsky JW, Ko C, Matskevitch AA, McKee HJ, Rossi JJ, Bouhamdan M, Pomerantz RJ, Kari G, Strayer DS (2004) Protecting from R5-tropic HIV: individual and combined effectiveness of a hammerhead ribozyme and a single-chain Fv antibody that targets CCR5. *Gene Ther* 11: 1627–1637
41 Steinberger P, Andris-Widhopf J, Buhler B, Torbett BE, Barbas CF 3rd (2000) Functional deletion of the CCR5 receptor by intracellular immunization produces cells that are refractory to CCR5-dependent HIV-1 infection and cell fusion. *Proc Natl Acad Sci USA* 97: 805–810
42 Reeves JD, Gallo SA, Ahmad N, Miamidian JL, Harvey PE, Sharron M, Pohlmann S, Sfakianos JN, Derdeyn CA, Blumenthal R et al. (2002) Sensitivity of HIV-1 to entry inhibitors correlates with envelope/coreceptor affinity, receptor density, and fusion kinetics. *Proc Natl Acad Sci USA* 99: 16249–16254
43 Nabatov AA, Pollakis G, Linnemann T, Kliphius A, Chalaby MI, Paxton WA (2004) Intrapatient alterations in the human immunodeficiency virus type 1 gp120 V1V2 and V3 regions differentially modulate coreceptor usage, virus inhibition by CC/CXC chemokines, soluble CD4, and the b12 and 2G12 monoclonal antibodies. *J Virol* 78: 524–530
44 Lee B, Sharron M, Blanpain C, Doranz BJ, Vakili J, Setoh P, Berg E, Liu G, Guy HR, Durell SR et al. (1999) Epitope mapping of CCR5 reveals multiple conformational states and distinct but overlapping structures involved in chemokine and coreceptor function. *J Biol Chem* 274: 9617–9626
45 Torre VS, Marozsan AJ, Albright JL, Collins KR, Hartley O, Offord RE, Quinones-Mateu ME, Arts EJ (2000) Variable sensitivity of CCR5-tropic human immunodeficiency virus type 1 isolates to inhibition by RANTES analogs. *J Virol* 74: 4868–4876
46 Rusert P, Kuster H, Joos B, Misselwitz B, Gujer C, Leemann C, Fischer M, Stiegler G, Katinger H, Olson WC et al. (2005) Virus isolates during acute and chronic human immunodeficiency virus type 1 infection show distinct patterns of sensitivity to entry inhibitors. *J Virol* 79: 8454–8469
47 Labrosse B, Labernardiere JL, Dam E, Trouplin V, Skrabal K, Clavel F, Mammano F (2003) Baseline susceptibility of primary human immunodeficiency virus type 1 to entry inhibitors. *J Virol* 77: 1610–1613
48 Scarlatti G, Tresoldi E, Bjorndal A, Fredriksson R, Colognesi C, Deng HK, Malnati MS, Plebani A, Siccardi AG, Littman DR et al. (1997) *In vivo* evolution of HIV-1 co-receptor usage and sensitivity to chemokine-mediated suppression. *Nat Med* 3: 1259–1265
49 Schmidtmayerova H, Sherry B, Bukrinsky M (1996) Chemokines and HIV replication. *Nature* 382: 767
50 Proudfoot AE, Power CA, Hoogewerf AJ, Montjovent MO, Borlat F, Offord RE, Wells TN (1996) Extension of recombinant human RANTES by the retention of the initiating methionine produces a potent antagonist. *J Biol Chem* 271: 2599–2603
51 Proudfoot AE, Buser R, Borlat F, Alouani S, Soler D, Offord RE, Schroder JM, Power CA, Wells TN (1999) Amino-terminally modified RANTES analogues demonstrate differential effects on RANTES receptors. *J Biol Chem* 274: 32478–32485
52 Townson JR, Graham GJ, Landau NR, Rasala B, Nibbs RJ (2000) Aminooxypentane addition to the chemokine macrophage inflammatory protein-1alpha P increases receptor affinities and HIV inhibition. *J Biol Chem* 275: 39254–39261
53 Sabbe R, Picchio GR, Pastore C, Chaloin O, Hartley O, Offord R, Mosier DE (2001) Donor- and ligand-dependent differences in C-C chemokine receptor 5 reexpression. *J Virol* 75: 661–671
54 Pastore C, Picchio GR, Galimi F, Fish R, Hartley O, Offord RE, Mosier DE (2003) Two mechanisms for human immunodeficiency virus type 1 inhibition by N-terminal modifications of RANTES. *Antimicrob Agents Chemother* 47: 509–517
55 Lederman MM, Veazey RS, Offord R, Mosier DE, Dufour J, Mefford M, Piatak M Jr, Lifson JD, Salkowitz JR, Rodriguez B et al. (2004) Prevention of vaginal SHIV transmission in rhesus macaques through inhibition of CCR5. *Science* 306: 485–487
56 Hartley O, Offord RE (2005) Engineering chemokines to develop optimized HIV inhibitors. *Curr Protein Pept Sci* 6: 207–219
57 Mosier DE, Picchio GR, Gulizia RJ, Sabbe R, Poignard P, Picard L, Offord RE, Thompson DA, Wilken J (1999) Highly potent RANTES analogues either prevent CCR5-using human immunod-

eficiency virus type 1 infection *in vivo* or rapidly select for CXCR4-using variants. *J Virol* 73: 3544–3550

58 Doranz BJ, Filion LG, Diaz-Mitoma F, Sitar DS, Sahai J, Baribaud F, Orsini MJ, Benovic JL, Cameron W, Doms RW (2001) Safe use of the CXCR4 inhibitor ALX40-4C in humans. *AIDS Res Hum Retroviruses* 17: 475–486

59 Doranz BJ, Grovit-Ferbas K, Sharron MP, Mao SH, Goetz MB, Daar ES, Doms RW, O'Brien WA (1997) A small-molecule inhibitor directed against the chemokine receptor CXCR4 prevents its use as an HIV-1 coreceptor. *J Exp Med* 186: 1395–1400

60 Tamamura H, Xu Y, Hattori T, Zhang X, Arakaki R, Kanbara K, Omagari A, Otaka A, Ibuka T, Yamamoto N et al. (1998) A low-molecular-weight inhibitor against the chemokine receptor CXCR4: a strong anti-HIV peptide T140. *Biochem Biophys Res Commun* 253: 877–882

61 Tamamura H, Hiramatsu K, Mizumoto M, Ueda S, Kusano S, Terakubo S, Akamatsu M, Yamamoto N, Trent JO, Wang Z et al. (2003) Enhancement of the T140-based pharmacophores leads to the development of more potent and bio-stable CXCR4 antagonists. *Org Biomol Chem* 1: 3663–3669

62 Olson W, Doshan H, Zhan C, Mezzatesta J, Assumma A, Czarnecky R, Stavola J, Maddon P, Kremer A, Israel R (2006) Prolonged coating of CCR5 lymphocytes by PRO 140, a humanized monoclonal antibody for HIV-1 therapy. 13th Conference on Retroviruses and Opportunistic Infections, Denver, CO, USA, 515

63 Labrijn AF, Poignard P, Raja A, Zwick MB, Delgado K, Franti M, Binley J, Vivona V, Grundner C, Huang CC et al. (2003) Access of antibody molecules to the conserved coreceptor binding site on glycoprotein gp120 is sterically restricted on primary human immunodeficiency virus type 1. *J Virol* 77: 10557–10565

64 Baba M, Nishimura O, Kanzaki N, Okamoto M, Sawada H, Iizawa Y, Shiraishi M, Aramaki Y, Okonogi K, Ogawa Y et al. (1999) A small-molecule, nonpeptide CCR5 antagonist with highly potent and selective anti-HIV-1 activity. *Proc Natl Acad Sci USA* 96: 5698–5703

65 Marozsan AJ, Kuhmann SE, Morgan T, Herrera C, Rivera-Troche E, Xu S, Baroudy BM, Strizki J, Moore JP (2005) Generation and properties of a human immunodeficiency virus type 1 isolate resistant to the small molecule CCR5 inhibitor, SCH-417690 (SCH-D). *Virology* 338: 182–199

66 De Clercq E, Yamamoto N, Pauwels R, Balzarini J, Witvrouw M, De Vreese K, Debyser Z, Rosenwirth B, Peichl P, Datema R et al. (1994) Highly potent and selective inhibition of human immunodeficiency virus by the bicyclam derivative JM3100. *Antimicrob Agents Chemother* 38: 668–674

67 Schols D, Vermeire K, Hatse S, Princen K, De Clercq E, Calandra G, Fricker S, Nelson K, Labrecque J, Bogucki D et al. (2004) *In vitro* anti-HIV activity profile of AMD887, a novel CCR5 antagonist, in combination with the CXCR4 inhibitor AMD070. 11th Conference on Retroviruses and Opportunistic Infections, San Francisco, CA, USA, 539

68 Ichiyama K, Yokoyama-Kumakura S, Tanaka Y, Tanaka R, Hirose K, Bannai K, Edamatsu T, Yanaka M, Niitani Y, Miyano-Kurosaki N et al. (2003) A duodenally absorbable CXC chemokine receptor 4 antagonist, KRH-1636, exhibits a potent and selective anti-HIV-1 activity. *Proc Natl Acad Sci USA* 100: 4185–4190

69 Tanaka Y, Okuma K, Tanaka R, Kumakura S, Shimoyamada A, Hirose K, Yanaka M, Murakami T, Yamamoto N (2006) Development of novel orally bioavailable CXCR4 antagonists, KRH-3955 and KRH-3140: binding specificity, pharmacokinetics and anti-HIV-1 activity *in vivo* and *in vitro*. 13th Conference on Retroviruses and Opportunistic Infections, Denver, CO, USA, 49LB

70 GlaxoSmithKline (2005) GlaxoSmithKline terminates patient enrollment for phase 3 studies of investigational HIV entry inhibitor aplaviroc (GW873140). http://www.gsk.com/media/pressreleases.htm. (November 8, 2005)

71 Greaves W, Landovitz R, Fatkenheuer G, Hoffmann C, Antunes F, Angel J, Boparai N, Knepp D, Keung A, Dunkle L (2006) Late virologic breakthrough in treatment-naive patients on a regimen of combivir + vicriviroc. 13th Conference on Retroviruses and Opportunistic Infections, Denver, CO, USA, 161LB

72 Hendrix CW, Collier AC, Lederman MM, Schols D, Pollard RB, Brown S, Jackson JB, Coombs RW, Glesby MJ, Flexner CW et al. (2004) Safety, pharmacokinetics, and antiviral activity of AMD3100, a selective CXCR4 receptor inhibitor, in HIV-1 infection. *J Acquir Immune Defic Syndr* 37: 1253–1262

73 Deeks SG (2006) Challenges of developing R5 inhibitors in antiretroviral naive HIV-infected patients. *Lancet* 367: 711–713

74 Wilkin T, Su Z, Kuritzkes DR, Hughes M, Flexner CW, Gross R, Coakley E, Greaves W, Godfrey C, Gulick R (2006) Co-receptor tropism in patients screening for ACTG 5211, a phase 2 study of vicriviroc, a CCR5 inhibitor. 13th Conference on Retroviruses and Opportunistic Infections, Denver, CO, USA, 655

75 Lalezari J, Thompson M, Kumar P, Piliero P, Davey R, Patterson K, Shachoy-Clark A, Adkison K, Demarest J, Lou Y et al. (2005) Antiviral activity and safety of 873140, a novel CCR5 antagonist, during short-term monotherapy in HIV-infected adults. *AIDS* 19: 1443–1448

76 Veazey RS, Klasse PJ, Schader SM, Hu Q, Ketas TJ, Lu M, Marx PA, Dufour J, Colonno RJ, Shattock RJ et al. (2005) Protection of macaques from vaginal SHIV challenge by vaginally delivered inhibitors of virus-cell fusion. *Nature* 438: 99–102

77 Veazey RS, Springer MS, Marx PA, Dufour J, Klasse PJ, Moore JP (2005) Protection of macaques from vaginal SHIV challenge by an orally delivered CCR5 inhibitor. *Nat Med* 11: 1293–1294

78 Arenzana-Seisdedos F, Virelizier JL, Rousset D, Clark-Lewis I, Loetscher P, Moser B, Baggiolini M (1996) HIV blocked by chemokine antagonist. *Nature* 383: 400

79 Schols D, Proost P, Struyf S, Wuyts A, De Meester I, Scharpe S, Van Damme J, De Clercq E (1998) CD26-processed RANTES(3-68), but not intact RANTES, has potent anti-HIV- 1 activity. *Antiviral Res* 39: 175–187

80 Simmons G, Clapham PR, Picard L, Offord RE, Rosenkilde MM, Schwartz TW, Buser R, Wells TNC, Proudfoot AE (1997) Potent inhibition of HIV-1 infectivity in macrophages and lymphocytes by a novel CCR5 antagonist. *Science* 276: 276–279

81 Menten P, Struyf S, Schutyser E, Wuyts A, De Clercq E, Schols D, Proost P, Van Damme J (1999) The LD78beta isoform of MIP-1alpha is the most potent CCR5 agonist and HIV-1-inhibiting chemokine. *J Clin Invest* 104: R1–5

82 Yang OO, Swanberg SL, Lu Z, Dziejman M, McCoy J, Luster AD, Walker BD, Herrmann SH (1999) Enhanced inhibition of human immunodeficiency virus type 1 by Met-stromal-derived factor 1beta correlates with down-modulation of CXCR4. *J Virol* 73: 4582–4589

83 Hartley O, Dorgham K, Perez-Bercoff D, Cerini F, Heimann A, Gaertner H, Offord RE, Pancino G, Debre P, Gorochov G (2003) Human immunodeficiency virus type 1 entry inhibitors selected on living cells from a library of phage chemokines. *J Virol* 77: 6637–6644

84 Murakami T, Nakajima T, Koyanagi Y, Tachibana K, Fujii N, Tamamura H, Yoshida N, Waki M, Matsumoto A, Yoshie O et al. (1997) A small molecule CXCR4 inhibitor that blocks T cell line-tropic HIV-1 infection. *J Exp Med* 186: 1389–1393

85 Mack M, Pfirstinger J, Haas J, Nelson PJ, Kufer P, Riethmuller G, Schlondorff D (2005) Preferential targeting of CD4-CCR5 complexes with bifunctional inhibitors: a novel approach to block HIV-1 infection. *J Immunol* 175: 7586–7593

86 Roschke V, Clark S, Branco L, Kanakaraj P, Kaufman T, Yao X, Nardelli B, Shi Y, Cai W, Ullrich S et al. (2004) Characterization of a panel of novel monoclonal antibodies that specifically antagonize CCR5 and block HIV-1 entry. 44th Interscience Conference on Antimicrobial Agents and Chemotherapy, Washington, DC, USA, H-213

87 Hatse S, Princen K, De Clercq E, Rosenkilde MM, Schwartz TW, Hernandez-Abad PE, Skerlj RT, Bridger GJ, Schols D (2005) AMD3465, a monomacrocyclic CXCR4 antagonist and potent HIV entry inhibitor. *Biochem Pharmacol* 70: 752–761

88 Princen K, Hatse S, Vermeire K, Aquaro S, De Clercq E, Gerlach LO, Rosenkilde M, Schwartz TW, Skerlj R, Bridger G et al. (2004) Inhibition of human immunodeficiency virus replication by a dual CCR5/CXCR4 antagonist. *J Virol* 78: 12996–13006

89 Princen K, Schols D (2005) HIV chemokine receptor inhibitors as novel anti-HIV drugs. *Cytokine Growth Factor Rev* 16: 659–677

90 Cumming JG, Cooper AE, Grime K, Logan CJ, McLaughlin S, Oldfield J, Shaw JS, Tucker H, Winter J, Whittaker D (2005) Modulators of the human CCR5 receptor. Part 2: SAR of substituted 1-(3,3-diphenylpropyl)-piperidinyl phenylacetamides. *Bioorg Med Chem Lett* 15: 5012–5015

91 Maeda K, Nakata H, Koh Y, Miyakawa T, Ogata H, Takaoka Y, Shibayama S, Sagawa K, Fukushima D, Moravek J et al. (2004) Spirodiketopiperazine-based CCR5 inhibitor which preserves CC-chemokine/CCR5 interactions and exerts potent activity against R5 human immunodeficiency virus type 1 *in vitro*. *J Virol* 78: 8654–8662

92 Dorr P, Westby M, Dobbs S, Griffin P, Irvine B, Macartney M, Mori J, Rickett G, Smith-Burchnell C, Napier C et al. (2005) Maraviroc (UK-427,857), a potent, orally bioavailable, and selective small-molecule inhibitor of chemokine receptor CCR5 with broad-spectrum anti-human immunodeficiency virus type 1 activity. *Antimicrob Agents Chemother* 49: 4721–4732

93 Strizki JM, Xu S, Wagner NE, Wojcik L, Liu J, Hou Y, Endres M, Palani A, Shapiro S, Clader JW et al. (2001) SCH-C (SCH 351125), an orally bioavailable, small molecule antagonist of the chemokine receptor CCR5, is a potent inhibitor of HIV-1 infection *in vitro* and *in vivo*. *Proc Natl Acad Sci USA* 98: 12718–12723

94 Strizki JM, Tremblay C, Xu S, Wojcik L, Wagner N, Gonsiorek W, Hipkin RW, Chou CC, Pugliese-Sivo C, Xiao Y et al. (2005) Discovery and characterization of vicriviroc (SCH 417690), a CCR5 antagonist with potent activity against human immunodeficiency virus type 1. *Antimicrob Agents Chemother* 49: 4911–4919

95 Takashima K, Miyake H, Kanzaki N, Tagawa Y, Wang X, Sugihara Y, Iizawa Y, Baba M (2005) Highly potent inhibition of human immunodeficiency virus type 1 replication by TAK-220, an orally bioavailable small-molecule CCR5 antagonist. *Antimicrob Agents Chemother* 49: 3474–3482

96 Baba M, Takashima K, Miyake H, Kanzaki N, Teshima K, Wang X, Shiraishi M, Iizawa Y (2005) TAK-652 inhibits CCR5-mediated human immunodeficiency virus type 1 infection *in vitro* and has favorable pharmacokinetics in humans. *Antimicrob Agents Chemother* 49: 4584–4591

97 Kwong PD, Wyatt R, Robinson J, Sweet RW, Sodroski J, Hendrickson WA (1998) Structure of an HIV gp120 envelope glycoprotein in complex with the CD4 receptor and a neutralizing human antibody. *Nature* 393: 648–659

98 Huang CC, Tang M, Zhang MY, Majeed S, Montabana E, Stanfield RL, Dimitrov DS, Korber B, Sodroski J, Wilson IA et al. (2005) Structure of a V3-containing HIV-1 gp120 core. *Science* 310: 1025–1028

99 Xu Y, Liu H, Niu C, Luo C, Luo X, Shen J, Chen K, Jiang H (2004) Molecular docking and 3D QSAR studies on 1-amino-2-phenyl-4-(piperidin-1-yl)-butanes based on the structural modeling of human CCR5 receptor. *Bioorg Med Chem* 12: 6193–6208

100 Fatkenheuer G, Pozniak AL, Johnson MA, Plettenberg A, Staszewski S, Hoepelman AI, Saag MS, Goebel FD, Rockstroh JK, Dezube BJ et al. (2005) Efficacy of short-term monotherapy with maraviroc, a new CCR5 antagonist, in patients infected with HIV-1. *Nat Med* 11: 1170–1172

Entry Inhibitors in HIV Therapy
Edited by Jacqueline D. Reeves and Cynthia A. Derdeyn

Inhibitors that target fusion

Wei Wang and Carol D. Weiss

Food and Drug Administration, Center for Biologics Evaluation and Research, Bldg. 29, Room 532, 29 Lincoln Drive, Bethesda, MD 20892-4555, USA

Introduction

The process of viral entry offers several advantages for drug intervention. Targeting extracellular events eliminates challenges in ensuring adequate drug delivery into cells. Disabling HIV before integration of viral DNA into host cells also prevents the potential for establishment of viral persistence in long-lived cells. Recent progress in understanding the molecular basis of the HIV entry process points to new therapeutic strategies. In this chapter, we focus on agents that target the step of virus-cell fusion from a mechanistic point of view. Issues related to product development of fusion inhibitors for eventual clinical use are covered in the Chapter by Greenberg.

Conformational changes leading to fusion

HIV enters cells through a multi-step process (Fig. 1, also reviewed in the Chapter by Tilton/Doms). Binding of gp120 to the CD4 receptor triggers conformational changes that facilitate gp120 interactions with a coreceptor (chemokine receptors CCR5 or CXCR4) [1, 2]. This interaction induces further conformational changes in the oligomeric envelope glycoprotein (Env) complex that activate the membrane fusion activity of gp41. gp41 draws viral and cellular membranes together as it refolds from its native, metastable structure to its final, thermostable structure (reviewed in [3]), leading to fusion of the outer leaflets of the membranes (hemifusion) and then complete membrane fusion with fusion pore formation (reviewed in [4]). Widening of the fusion pore allows the viral nucleocapsid to be delivered into the host cell.

The gp41 ectodomain (Fig. 2A) contains five essential regions for fusion: a fusion peptide (FP), two heptad repeats (HR1 and HR2), a membrane proximal region (MPR) and a transmembrane domain (TM). The FP, containing predominantly hydrophobic residues, begins at the extreme N terminus of the gp41 ectodomain and precedes HR1 (amino acid positions 29–82), a predicted -helical, coiled-coil domain, also called the N heptad or N peptide region. An intervening sequence, which contains a small loop formed by an intramol-

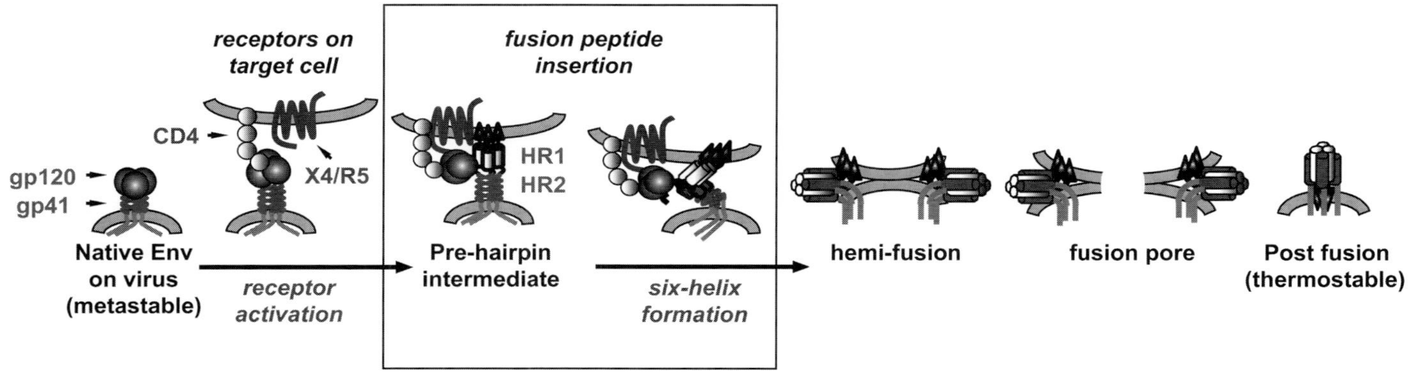

Figure 1. Multi-step process of HIV entry. Binding of the native, metastable envelope glycoprotein (gp120/41) to CD4 and chemokine cellular receptors (X4 or R5) activates conformational changes that lead to fusion. The changes involve opening up of gp41 to form the pre-hairpin fusion intermediate, in which gp41 becomes embedded in both viral and target membranes and the gp41 heptad repeats (HR1 and HR2) are relatively exposed. Further folding to form the compact, thermostable six-helix structure facilitates merger of viral and target membranes. Current fusion inhibitors target the pre-hairpin fusion intermediate conformation(s) of the envelope glycoprotein (inside box).

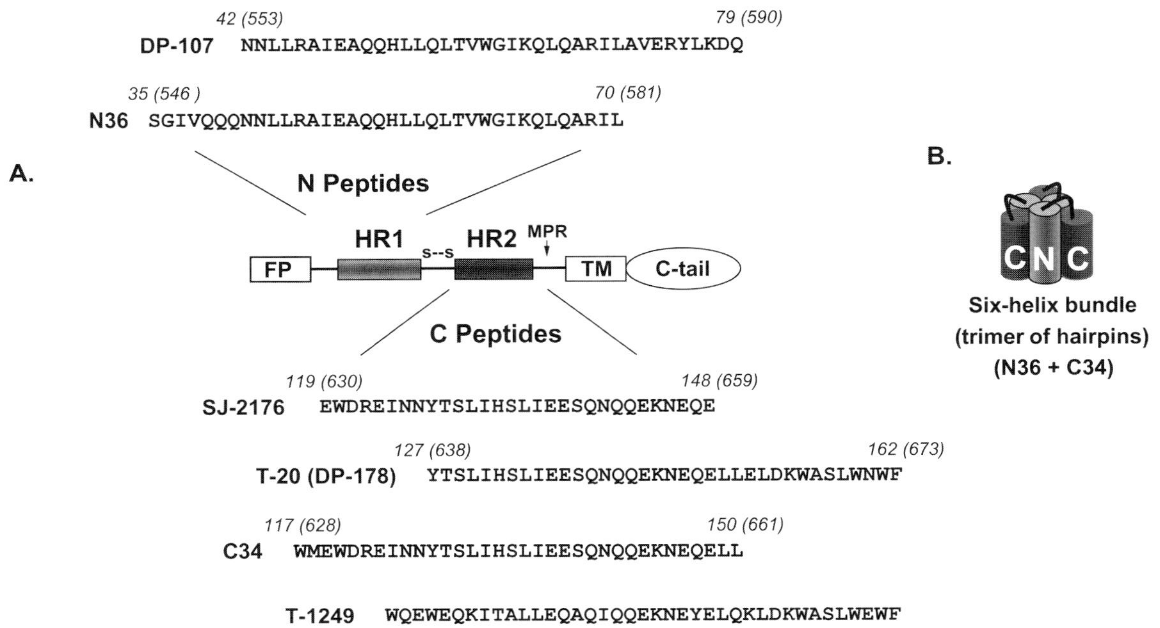

Figure 2. Diagram of gp41 ectodomain. (A) gp41 domains and corresponding peptides are notated. FP is fusion peptide. HR1 and HR2 are the heptad repeat motifs. MPR is the membrane proximal region. TM is transmembrane domain. C-tail is cytoplasmic tail. Numbers denote amino acid starting at the N terminus of gp41 or, in parentheses, starting at gp160 according to the nomenclature of the HXB2 reference clone. (B) Schematic of six-helix bundle. HR1 and HR2 segments (corresponding N36 and C34 peptides, respectively) self-assemble into a stable helical bundle. C peptides pack in an anti-parallel orientation against the N peptide, trimeric coiled-coil core.

ecular cysteine bridge within each gp41 subunit, separates the HR1 and HR2. The HR2 (amino acid positions 117–162), also referred to as C heptad or C peptide region, follows the loop region and is contiguous with a tryptophan-rich, MPR. The MPR and C-terminal part of the HR2 are notable targets for rare, broadly neutralizing antibodies to HIV [5, 6], which interfere with virus-cell fusion. Finally, the TM anchors Env in the membrane through its stretch of hydrophobic residues.

Receptor engagement triggers the fusion process by releasing gp41 from its native, non-fusogenic conformation, so that it undergoes fusogenic conformational changes as it folds into the stable, six-helix conformation (also referred to as a trimer of hairpins). The six-helix structure (Fig. 2B), identified from a protease-resistant fragment of gp41, consists of HR1 and HR2 peptides that self-assemble into a stable helical bundle [7]. The six-helix bundle has a trimeric, coiled-coil core, comprised of three HR1 segments, against which three HR2 helices pack in an anti-parallel fashion into the hydrophobic grooves of the coiled coil [8–10]. During the initial stages of Env refolding, the FP is believed to relocate from the molecular interior of Env so that it can insert in the target cell membrane, in a manner similar to the spring-loaded mechanism described for the well-characterized, influenza hemagglutinin [11]. In this "sprung" conformation, referred to as pre-hairpin fusion intermediate, gp41 bridges the cell and viral membranes through its FP and the transmembrane domains, respectively. Subsequent folding of the HR1 and HR2 into the six-helix bundle promotes fusion by bringing viral and cellular membranes in close apposition and releasing energy as gp41 assumes its more stable conformation [12].

Opportunities and challenges for inhibiting gp41-mediated fusion

The fusion process offers opportunities for developing potent, virus-specific inhibitors. A fundamental step in fusion, with demonstrated potential for therapeutic intervention, is the interaction between HR1 and HR2 to form the six-helix bundle. Six-helix formation drives fusion and agents that prevent its formation block HIV entry. Prior to the attainment of the six-helix structure, the transient pre-hairpin intermediate structure(s) increases exposure of the HR1 and HR2 [12–14], and thus opens up at least two sites on gp41 that can be targeted by inhibitors (Fig. 3). While many aspects of the fusion process are still poorly understood, for the purposes of this chapter any agent that interferes with fusion after receptor binding will be referred to as a fusion inhibitor. Thus far, fusion inhibitor development has focused primarily on peptides or other molecules that interfere with HR1 and HR2 interactions.

The first examples of HIV fusion inhibitors were discovered in the early 1990s as research peptides that prevented HIV infection in laboratory cultures [15–17]. Indeed, the discoveries that gp41 peptides mimicking HR1 and HR2 could block HIV infection, along with subsequent high-resolution studies of

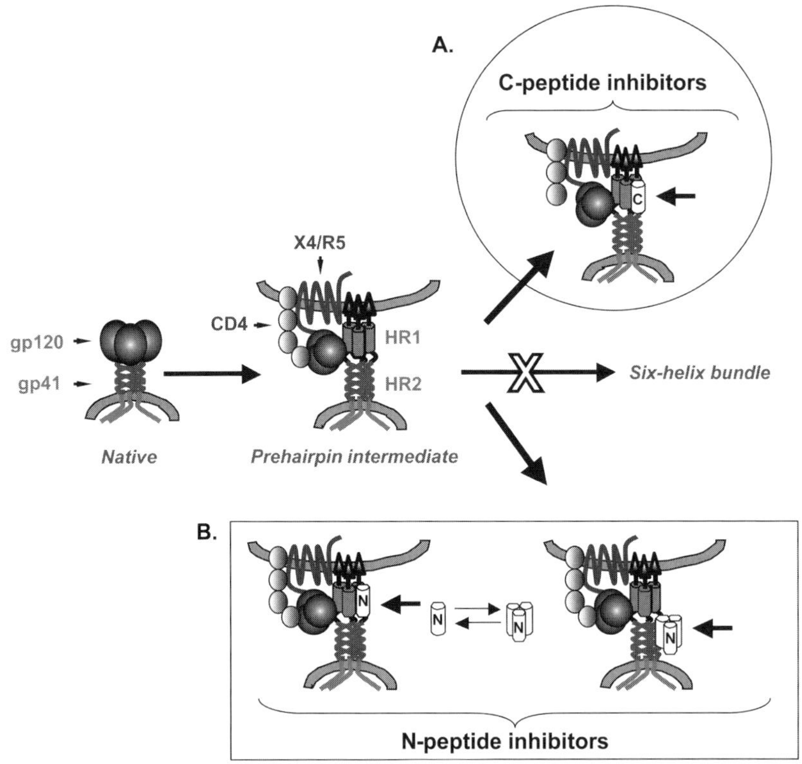

Figure 3. Model of dominant-negative mechanism of inhibition by peptide fusion inhibitors. (A) C peptides bind the gp41 HR1 to form a peptide-gp41 six-helix that cannot mediate membrane fusion. (B) N peptide fusion inhibitors may bind the HR2 to form a peptide-gp41 six-helix bundle, or the HR1 to form a peptide-gp41 coiled coil. Peptide binding to gp41 traps the fusion intermediate and interferes with bundle formation by the endogenous gp41 HRs.

the gp41 self-assembly domain, helped to elucidate the current models of HIV entry. The strong self-association properties of the HR1 and HR2 peptides, resulting in formation of the six-helix bundle, make it likely that these inhibitory peptides bind gp41 HRs in a similar manner to form a heterologous peptide-gp41 bundle [18] (Fig. 3). Peptide binding to gp41 consequently interrupts formation of the six-helix bundle by the endogenous gp41 HRs [19, 20]. According to this so-called dominant-negative mode of inhibition, peptides corresponding to the HR2 (also called C peptides) operate by binding along the hydrophobic grooves of the HR1 trimeric, coiled-coil core (Fig. 3A). In an analogous manner, peptides corresponding to the HR1 (also called N peptides), which spontaneously form coiled-coil structures, bind to the HR2 [14] (Fig. 3B). However, N peptides may also bind the HR1 to form a heterologous peptide-gp41 coiled coil [21], which interferes with the endogenous coiled coil and prevents formation of the gp41 six-helix bundle (Fig. 3B). That the N and C peptides bind different residues and sites on gp41 raises the possibility that they may potentially represent different subclasses of fusion inhibitors with different resistance profiles, as discussed further below. Non-peptidic inhibitors that target HR1 and HR2 are also being developed (reviewed in [22, 23]).

While fusion inhibitors offer great promise as a potent new class of antiretrovirals, there are theoretical and practical constraints that pose challenges for developing new agents in this class. Fusion inhibitors, according to our operational definition, target transient Env structures that occur after receptor binding and before formation of the six-helix bundle. Consequently, both kinetic and structural (steric) restrictions may limit potency of potential inhibitors. In the first case, the kinetics of virus-cell fusion is such that fusion inhibitors have a relatively short time frame to bind fusion intermediate structures, compared to drugs that target other, long-lived structures, such as stable cell receptors. Targeting Env structures after receptor engagement also means that inhibitors may have limited access between the closely apposed viral and cellular membranes. gp41 may be further occluded by gp120 and the cellular receptors. In other words, fusion inhibitors must catch a short-lived, moving target in a tight space.

There are also inherent challenges to blocking HR1 and HR2 interactions. The long helices of the HR1 make extensive contacts with each other and with the long helices of the HR2 that pack into the hydrophobic coiled-coil grooves, thus making the six-helix bundle an unusually stable structure. Therefore, it may be difficult for a small molecule with few contacts with gp41 to prevent association between the HR1 and HR2. To date, the most potent fusion inhibitors are in fact peptides that have potential to make extensive gp41 contacts. Yet it would seem that even peptide inhibitors are surprisingly potent, because it is hard to explain how the N and C peptides can effectively compete with the endogenous HR1 and HR2 of gp41 to form the six-helix bundle. The endogenous HR1 and HR2 only have to undergo unimolecular interactions and should have an advantage in competing with the exogenous N and C peptides that must undergo bimolecular interactions. Therefore, it is widely assumed that the N and C peptides bind the pre-hairpin intermediate at a time when the endogenous HRs are still restricted from folding into the complete six-helix structure. In this case, small molecules with access to the intermediate would also have an advantage over the endogenous HR. So it appears that the kinetics of Env refolding also provide opportunities for developing inhibitors.

Other issues pertaining to large-scale manufacture and poor oral availability of the peptides raise technical challenges for fusion inhibitor development that are discussed in the Chapter by Greenberg. Finally, aside from the opportunities and challenges that are specific to fusion inhibitors, all HIV inhibitors unfortunately face the problem of emerging viral resistance. It will therefore be important to develop fusion inhibitors that bind gp41 in different ways to offset potential for cross-resistance among agents in the fusion inhibitor class.

C-peptide fusion inhibitors

Synthetic peptides based on the HR2 amino acid sequence, also called C peptide fusion inhibitors, have proven to be potent inhibitors of HIV infection both *in vitro* and *in vivo* [24, 25]. The prototype in this class is the peptide called DP-178 [17], now widely referred to as T-20 or by the drug name Enfuvirtide/Fuzeon[R]. This 36-residue peptide encompasses most of the C-terminal HR2 sequence (Fig. 2 and Tab. 1) and overlaps other HR2 peptides, including SJ-2176 [16] and C34 [26], which also potently inhibit HIV infection *in vitro*. Structural studies of HR1 and HR2 peptides (N36 and C34) that self-assemble into a thermostable, six-helix bundle [8] predict that T-20 binds the HR1 along the coiled coil grooves, and thus prevents bundle formation according to the dominant negative mechanism discussed above. This model is supported by a wide variety of data. Genetic studies show that viruses grown in the presence of T-20 frequently generate mutations in the N-terminal portion of HR1 that reduce sensitivity to the peptide [27, 28]. Biochemical studies with peptides and recombinant proteins further show that C peptides can bind gp41 [13, 29–31] and that HR1 mutations can impair association with HR2 peptides [20, 27, 32–34]. The stability of the bundle formed by T-20 and HR1 also correlates with the inhibitory activity of T20 [35, 36].

Table 1. Fusion inhibitor table

Inhibitor	Type	Major target	~ Potency[a] (*in vitro*)
T-20/Enfuvirtide[R] (DP-178) [24, 25]	C peptide	HR1	nM
T-649 (C34) [26, 39]	C peptide	HR1	nM
T-1249 [93]	C peptide	HR1	nM
D10-p5-2K, D10-p3-2K [53]	D peptide version of C peptide	HR1 pocket	μM
C14linkmid [50]	C peptide	HR1	μM
HIV31 [48]	C peptide	HR1	nM
N34(L6)C28 [94]	Recombinant protein	HR1 (?)	nM
ADS-J1, ADS-J2 [55, 57, 58]	Small molecule	HR1 pocket	μM
XTT formazan [59]	Small molecule	HR1 pocket	μM
T-21 (DP-107) [15]	N peptide	HR2, HR1	nM- μM
N36 [64]	N peptide	HR2, HR1	nM- μM
IZN36 [64]	N peptide	HR2	nM
5-Helix [67]	Recombinant protein	HR2	nM
N_{CCG}-gp41 [68]	Recombinant protein	HR2	nM
$N36^{mut(e,g)}$ [21]	N peptide	HR1	>μM

[a] Inhibitor potency is given as the approximate 50% inhibitory concentration in nM or μM ranges in standard *in vitro* infectivity assays.

Yet many of the molecular details of the mechanism of T-20 inhibition are still not known. Although it is clear that T-20 interactions with the HR1 are critical to its inhibitory activity, there are data indicating that T-20 potentially interacts with other regions of Env. It has been suggested that the C-terminal portion of T-20, which is not part of the six-helix structure revealed in high-resolution studies, may interact with the membrane or regions of gp41 that are near or within the membrane [37–39]. Two studies also provide evidence that T-20 binds near the co-receptor binding site on gp120 in viruses that use the CXCR4 co-receptor, possibly involving electrostatic interactions [40, 41]. The precise contributions of T-20 binding to non-HR1 regions of Env to the inhibitory activity of T-20 need further study and may yield new targets for drug development.

Other C peptides are also potent inhibitors of HIV fusion (Fig. 2, numbers correspond to the Los Alamos numbering for the reference HXB2 clone shown as residue of gp41 and in parentheses as residue of gp160). C34, corresponding to HR2 residues that completely overlap the HR1 helices found in the six-helix structure, forms a more stable six-helix bundle with N peptides compared to T-20 [39, 42]. Only the N-terminal two-thirds of T-20 overlap HR1 in the bundle. Although the C34 and T-20 peptides both inhibit HIV infection in the nanomolar range [26], the N-terminal residues of C34, in particular W117 (W628), W120 (W631), and I124 (I635), provide extra contacts with residues in a conserved hydrophobic pocket in the C-terminal part of the HR1 coiled coil. These contacts play a role in strengthening anti-HIV activity [26]. For this reason, C peptide inhibitors containing these pocket-binding residues, such as T-649 (same sequence as C34 except with a two-residue extension at the C terminus) and T-1249, have been considered for clinical development (see Chapter by Greenberg). That T-649 was found to be more potent than T-20 against a panel of 55 primary isolates [43] and that resistance to C34 was found to develop more slowly than resistance to T-20 [44] are consistent with biophysical studies showing that contacts in the hydrophobic pocket stabilize the six-helix structure [26]. T-1249 is a synthetic, designed peptide that has characteristics of both T-20 and T-649 [45] and is effective against T-20-resistant viruses [46, 47]. However, problems relating to the formulation of T-649 and T-1249 have refocused efforts on newer generation C peptides [45] (www.trimeris.com). Because C peptides containing the hydrophobic pocket-binding residues are often less soluble than T-20, amino acid substitutions or other modifications to non-critical residues of the peptide may be introduced to improve solubility. Approaches for modifying C peptides have included substitutions with helix-favoring amino acids and covalent cross-linkers that stabilize the helices [48–51]. Such modifications allow the use of smaller peptides without compromising antiviral potency, although there is not always a strict correlation between helical propensity and inhibitory activity [50].

Small-molecule fusion inhibitors

Intensive searches for small compounds with C peptide-like inhibitory activity are also underway (reviewed in [22, 23]). In these efforts, the HR1 pocket has been an attractive target for rational drug design [26], because small molecules that dock into the pocket could prevent formation of the six-helix bundle required for fusion. In addition, mutational analyses show that residues L57 (L568) and W60 (W571) in the conserved hydrophobic pocket are critical for membrane fusion activity [52] and would therefore probably not tolerate mutations that could lead to resistance. Strategies for identifying pocket-binding compounds often include the use of engineered, HR1 coiled-coil peptides that stably present the hydrophobic pocket to potential inhibitors. In one approach, mirror-image phage display technology was used to identify cyclic D-amino acid peptides that inhibit HIV entry [53]. In another approach, biased combinatorial libraries of small molecules attached to short segments of HR2 peptides were used to select molecules that fit into the hydrophobic pocket, although these small molecules alone did not exhibit potent inhibitory activity [54]. Other strategies emphasize molecular modeling to virtually screen libraries of compounds that would likely dock into the hydrophobic pocket or otherwise interfere with six-helix bundle formation. This approach led to the discovery of ADS-J1 and ADS-J2, lead compounds that inhibit HIV infection in the micromolar range and block HR1 and HR2 interactions in a peptide-based assay [55–57], although other inhibitory mechanisms may also be involved [58]. High-throughput screens have also identified other small compounds with anti-HIV activity, including XTT formazan [59], NB-2, NB-64 [60], RPR103611 [61], KY001 and KY002 [62]. Further efforts in these areas promise to yield more small-molecule inhibitors with improved bioavailability and more favorable drug-like properties [63].

N peptide fusion inhibitors

In vitro studies indicate that the HR2 region of gp41 is also a viable target for inhibitors. Several peptides mimicking the HR1, referred to as N peptide fusion inhibitors, potently inhibit laboratory cultures of HIV. The first example, reported in the early 1990s, is a 38-amino acid peptide called DP-107 (sequence similar to T-21) [15]. The DP-107 peptide forms coiled-coil structures in solution, and its inhibitory activity is correlated with its ability to adopt helical structure [15]. The N36 peptide, corresponding to the 36 residues in HR1 that overlap DP-107 and comprise the coiled-coil core of the six-helix, is also a potent inhibitor. As described above, N peptides spontaneously assemble with C peptides to form six-helix-like structures. N peptide inhibitory potency likely depends on interactions with HR2 to form a peptide-gp41, six-helix structure that prevents formation of the endogenous six-helix bundle [64]. This interaction with the HR2 depends on the conserved, hydrophobic residues in the grooves of the HR1 coiled coil. Accordingly, the hydrophobic residues of the C peptides line up on one face of the peptide as the HR2 assumes the helical structure needed to interact with the HR1 coiled-coil grooves. In the absence of C peptide interactions, the N peptide coiled-coil trimer is much less stable and tends to form higher order oligomeric structures and aggregates [65], as it tries to shield its hydrophobic residues from the aqueous environment. This property effectively reduces the concentration of the trimeric species, which is thought to be responsible for most of the inhibitory activity [64].

Efforts to improve solubility and stabilize the N coiled-coil trimer using a rational design have yielded a variety of constructs with dramatically improved inhibitory potency compared to DP-107 [64, 66]. Examples of potent N peptide inhibitors designed to stabilize the trimeric coiled coil include IQN17 [53], IZN36 [64], 5-Helix [67], and $N36_{CCG}$-gp41 [68], which inhibit HIV cultures in the low nanomolar range similar to T-20. The first two examples are chimeric synthetic peptides made with variable lengths of the HR1 fused to unrelated soluble, trimeric coiled coils. Other examples take advantage of the self-assembly properties of the HR1 and HR2 segments to stabilize the N coiled-coil trimer. The 5-Helix inhibitor is essentially a six-helix recombinant protein that lacks one of the HR2 helices [67], thereby exposing one groove of the coiled coil for capturing a gp41 HR2 segment. The Pseudomonas exotoxin protein has also been attached to the 5-Helix for selective cell killing of Env-expressing cells [30]. Another recombinant protein, $N36_{CCG}$-gp41 [68], uses a complete six-helix bundle as a stabilizing base to project a duplicate HR1 coiled-coil trimer extending from the coiled-coil core of the six-helix bundle. This HR1 extension is further stabilized by the introduction of cysteine residues that form covalent bonds between the helices of the extended HR1 coiled coil.

N peptides also have the potential to interact with the HR1 segment of gp41, according to the dominant-negative mechanism described above. Acting as monomers or dimers instead of trimers, N peptides could interact with the endogenous HR1 segments in gp41 to form a peptide-gp41 coiled coil that would interfere with gp41 conformational changes (Fig. 3B). This possibility is supported by experiments with a synthetic N peptide inhibitor, $N36^{mut\,(e,g)}$ [21], in which residues were mutated to prevent HR2 interactions. This peptide demonstrates significant inhibitory activity in a modified fusion assay that is unusually sensitive to inhibitors, although it is much less potent in traditional fusion assays. Nonetheless, the demonstration of inhibitory activity with $N36^{mut\,(e,g)}$ suggests that N peptides may target both sites of gp41, perhaps with different efficiencies.

As with C peptides, the N peptides and similar recombinant proteins face hurdles related to poor oral availability and difficulties in discovering small-molecule mimics of HR1 that would be capable of interfering with protein-protein interactions in the endogenous, six-helix bundle. Additionally, while protein engineering offers good opportunities for creating constructs with the desired stability and solubility, larger inhibitors become increasingly immunogenic and vulnerable to degradation and immune clearance. Yet, development of N peptide-like inhibitors has the potential to offer an important therapeutic advance. Because N and C peptides target different sites on gp41, they potentially represent different subclasses of fusion inhibitors with different resistance profiles.

Resistance to fusion inhibitors

C peptide resistance

All current antiretrovirals, including T-20, inhibit spread of HIV infection to new target cells, but they do not kill HIV-infected cells. Therefore, any reduction of drug pressure results in rapid return of progeny virus from infected cells. To maintain maximal viral suppression, combination chemotherapy is needed to exert a high genetic barrier to resistance.

Many data on the viral determinants of resistance to T-20 have accumulated from clinical and laboratory studies, and are covered in detail in the Chapter by Greenberg. Initially, *in vitro* selection studies identified mutations in the GIV sequence of gp41, corresponding to residues 36–38 of gp41 (547–549 of gp160) that conferred high-level resistance to T-20 in culture [27]. For example, the G36S/V38M mutations show about a 100-fold decrease in susceptibility to T-20 [27]. Subsequently, mutations in the same residues were found to emerge in patients who developed clinical resistance while on T-20 therapy [28, 69]. Neighboring, C-terminal gp41 residues 36–45 (547–556 of gp160), GIVQQQNNLL within HR1, are also often mutated in patients resistant to T-20, including residues V38, Q40, N42, N43 and L45 [46, 70–72]. Single mutations can cause a 5–10-fold loss of susceptibility to T-20, while double mutations, which are more frequently seen, can lead to greater levels of resistance (reviewed in [70]).

Mutations in the HR2 also arise during T-20 treatment, often subsequent to HR1 mutations [46, 72–74]. Studies involving computational protein modeling further suggest that HR2 mutations can compensate for a loss of bundle stability that often results from primary resistance mutations in HR1 [75]. Such compensatory HR2 mutations are believed to increase six-helix stability, resulting in greater resistance and viral fitness [75]. However, resistance mutations typically revert to wild type after cessation of therapy [76], consistent with the notion that these mutations impair viral fitness. *In vitro* assays indicate that common resistance mutations in the HR1 often have fitness costs [46, 47, 77–79], but examples of resistant viruses with apparent high viral fitness can also be found [78, 79].

Collectively, the literature implicates several mechanisms for T-20 resistance. HR1 mutations are most common and appear to play a primary role in resistance [33]. Biophysical experiments suggest that some HR1 mutations, such as those in position 43, impair C peptide binding to the HR1 coiled-coil groove [27, 33, 34]. Most resistance mutations, however, seem to operate more indirectly, perhaps by favoring formation of the endogenous gp41 bundle over the heterologous bundle formed between gp41 and the peptide. Additionally, a correlation between fusion kinetics and sensitivity to T-20 has been demonstrated, leading to the proposal that faster fusion kinetics reduces sensitivity to T-20 [80]. In this scenario, the shortened half-life of the pre-hairpin intermediate reduces its availability for inhibitor binding. The increased kinetics could be due to the intrinsic properties of Env, such as higher receptor affinity due to mutations in gp120 [43, 80], or host cell factors such as higher density of receptors [80] or adhesion molecules [81]. Reduced receptor binding, however, does not always lead to increased sensitivity to a fusion inhibitor [82]. Given the complex interactions involved in virus entry, it may therefore be difficult to predict sensitivity to peptide fusion inhibitors based on viral genetics alone [83], and it is likely that several factors contribute to viral susceptibility to peptide fusion inhibitors. Nonetheless, emergence of key mutations, such as V38A, may be a common step in several resistance pathways.

N peptide resistance

Since N peptides have not entered the clinic, resistance data are limited. One report on a resistance pathway to an N peptide inhibitor (N44*) corresponding to a 44-residue peptide that overlaps DP-107 and N36 shows interesting similarities with resistance to C peptides [84]. *In vitro* selection of an HIV molecular clone showed that an initial mutation E137K (E648K) arose in HR2, followed by a second mutation Q66R (Q577R) in HR1, reminiscent of the sequential HR mutations arising during T-20 selection, only in the reverse order [34, 73]. The resistance mutations conferred a slightly higher level of resistance to other N peptide inhibitors with stabilized coiled coils, including 5-Helix and IZN36. Unexpectedly, both mutations also conferred resistance to C peptide inhibitors, including T-20 and C34, at a level similar to the selecting N peptide. Biophysical studies comparing bundles formed by mutant or wild-type N and C peptides indicate that the mutations would increase the bundle stability [84], thus favoring the endogenous bundle over the bundle formed with the peptide. The N peptide resistant virus also demonstrated increased susceptibility to soluble CD4, suggesting that resistance also may involve changes in receptor activation and fusion kinetics. In a separate preliminary report, a virus selected for resistance to the 5-Helix also developed mutations in HR1 and HR2 [85]. These initial reports of N peptide resistant viruses alert us to the possibility that there will be mechanistic similarities in resistance pathways for N and C peptides and that some pathways could lead to cross-resistance among fusion inhibitors. More complete descriptions of the resistance pathways for N peptides and how they relate to resistance pathways for C peptides are needed for informed development of new fusion inhibitors.

Natural susceptibility to fusion inhibitors

Variability of the HR1 and HR2

The high degree of genetic variability in Env and the multi-step process of entry result in a wide range of sensitivities to fusion inhibitors among HIV strains. At the same time, the conserved nature of the HRs predicts that these regions would be good targets for inhibitors [86]. This prediction has been borne out by the success of T-20 as a potent arm in combination antiretroviral regimens, despite wide-ranging sensitivities of primary viruses to peptide fusion inhibitors [87, 88].

There are at least nine circulating subtypes and recombinant forms of HIV within the prevalent M group. Among the subtypes, the HR1 region is one of the most highly conserved regions (reviewed in [89]). Variations near the GIV motif and in the hydrophobic pocket are found only infrequently. This high degree of conservation is likely due to the extensive protein-protein interactions that occur in this region. Many HR1 residues are involved not only in coiled-coil interactions within the HR1 trimer, but also in interactions with the HR2 to form the six-helix structure. Within the HR1, the residues that typically contribute most to coiled-coil interactions reside in the "*a*" and "*d*" positions in the "*abcdefg*" heptad repeat nomenclature, while residues in the "*e*" and "*g*" positions generally interact with the HR2 helices. Hydrophobic interactions provide major stabilizing forces within the six-helix bundle [52]. Aside from the extensive network of interactions between helices, variability of the HR1 may also be restricted by the need to conserve a functional Rev-responsive element (RRE). The RRE is an RNA hairpin structure residing within the HR1 region, which is important for transport of messenger RNA out of the nucleus. Selection to preserve a functional RRE was evident in a patient with T-20 resistance who developed two simultaneous substitutions located complementary to each other in the RRE [34]. These two substitutions maintained the secondary structure of the RRE that was impaired by a prior resistance mutation in HR1.

The HR2 is more variable relative to HR1, though it still has a high percentage of conserved residues. The highly conserved residues, mostly residing in the "*a*" and "*d*" positions of the HR2, align on the face of the helix that is predicted to interact with the grooves of the HR1 coiled coil. Some residues, including W117(628), E123(634), L134(645), Q142(653), N145(656), and L149(660) are invariant among the clade B and C subtypes. In contrast, the HR1 has 13 invariant residues among the same subtypes.

Co-receptor usage and sensitivity to T-20

Factors other than polymorphisms in HR1 and HR2 affect sensitivity to fusion inhibitors [43, 80, 90, 91]. Because fusion inhibitors target a structural intermediate, interactions with receptors that affect the half-life of the pre-hairpin fusion intermediate would likely influence the window of action for T-20 [80]. Several studies suggest that HIV envelope co-receptor tropism or affinity might contribute to the wide range of susceptibility to T-20 measured *in vitro*. In two panels of T-20-naïve, primary isolates, CCR5-using viruses were generally found to be less sensitive to T-20 compared to CXCR4-using viruses [43, 90]. The correlation also extended to T-649 [43]. Experiments with engineered Env chimeras further showed that this sensitivity mapped to the V3 loop of gp120, consistent with co-receptor usage influencing susceptibility to the inhibitor [43].

On the other hand, a different panel involving a large number of isolates from a clinical study showed no differences between CXCR4- or CCR5-using isolates in sensitivity to T-20 [69]. Furthermore, similar sensitivities to T-20 were also seen between different clones from longitudinal samples from a single patient who underwent a switch in co-receptor tropism [70]. Importantly, co-receptor usage at initiation of T-20 treatment could not be correlated with response to treatment, and patients harboring CCR5-tropic, CXCR4-tropic, or dual tropic viruses responded to T-20 treatment with equivalent reductions in viral load [92]. Nonetheless, *in vitro* studies clearly show that point mutations that reduce co-receptor affinity can increase susceptibility to fusion inhibitors [82], although the extent to which these *in vitro* findings will translate into clinically important differences remains unclear. Altogether, these studies underscore the complex interactions within Env and between Env and co-receptors that influence sensitivity to fusion inhibitors.

Conclusion

Fusion inhibitors represent an emerging new class of potent antiretroviral agents that are urgently needed to suppress HIV in the face of growing resistance to reverse transcriptase and protease inhibitors. Despite the rather unconventional mode of action against a transient viral target, fusion inhibitors are clinically effective and offer a new way to disable HIV. Agents in this class also have the inherent therapeutic advantages of targeting extracellular and highly conserved sites on the virus. Thus, peptide fusion inhibitors are generally potent against a wide range of HIV strains, and resistance often comes with a fitness cost. At the same time, the mode of action faces the hurdle of interrupting stable protein-protein interactions within Env, which cannot be easily achieved with small-molecule inhibitors. At present only peptides are highly potent, but these large molecules present their own challenges, especially relating to oral bioavailability. As the field moves forward with a more in-depth understanding of HIV fusion, new avenues for interventions will come into view, in part aided by the use of peptide fusion inhibitors as valuable probes for elucidating Env-conformational changes that could be targeted by novel inhibitors.

Acknowledgements
We gratefully acknowledge Nidhi Gupta and Russell Vassell for helpful discussions and critical reading of the manuscript.

References

1 Trkola A, Dragic T, Arthos J, Binley JM, Olson WC, Allaway GP, Cheng-Mayer C, Robinson J, Maddon PJ, Moore JP (1996) CD4-dependent, antibody-sensitive interactions between HIV-1 and its co-receptor CCR-5. *Nature* 384: 184–187
2 Wu L, Gerard NP, Wyatt R, Choe H, Parolin C, Ruffing N, Borsetti A, Cardoso AA, Desjardin E, Newman W et al. (1996) CD4-induced interaction of primary HIV-1 gp120 glycoproteins with the chemokine receptor CCR-5. *Nature* 384: 179–183
3 Eckert DM, Kim PS (2001) Mechanisms of viral membrane fusion and its inhibition. *Annu Rev Biochem* 70: 777–810
4 Gallo SA, Finnegan CM, Viard M, Raviv Y, Dimitrov A, Rawat SS, Puri A, Durell S, Blumenthal R (2003) The HIV Env-mediated fusion reaction. *Biochim Biophys Acta* 1614: 36–50
5 Muster T, Guinea R, Trkola A, Purtscher M, Klima A, Steindl F, Palese P, Katinger H (1994) Cross-neutralizing activity against divergent human immunodeficiency virus type 1 isolates induced by the gp41 sequence ELDKWAS. *J Virol* 68: 4031–4034
6 Zwick MB, Labrijn AF, Wang M, Spenlehauer C, Saphire EO, Binley JM, Moore JP, Stiegler G, Katinger H, Burton DR, Parren PW (2001) Broadly neutralizing antibodies targeted to the membrane-proximal external region of human immunodeficiency virus type 1 glycoprotein gp41. *J Virol* 75: 10892–10905
7 Lu M, Blacklow SC, Kim PS (1995) A trimeric structural domain of the HIV-1 transmembrane glycoprotein. *Nat Struct Biol* 2: 1075–1082
8 Chan DC, Fass D, Berger JM, Kim PS (1997) Core structure of gp41 from the HIV envelope glycoprotein. *Cell* 89: 263–273
9 Weissenhorn W, Dessen A, Harrison SC, Skehel JJ, Wiley DC (1997) Atomic structure of the ectodomain from HIV-1 gp41. *Nature* 387: 426–430
10 Tan K, Liu J, Wang J, Shen S, Lu M (1997) Atomic structure of a thermostable subdomain of HIV-1 gp41. *Proc Natl Acad Sci USA* 94: 12303–12308
11 Carr CM, Kim PS (1993) A spring-loaded mechanism for the conformational changes of influenza hemagglutinin. *Cell* 73: 823–832
12 Melikyan GB, Markosyan RM, Hemmati H, Delmedico MK, Lambert DM, Cohen FS (2000) Evidence that the transition of HIV-1 gp41 into a six-helix bundle, not the bundle configuration, induces membrane fusion. *J Cell Biol* 151: 413–424
13 Furuta RA, Wild CT, Weng Y, Weiss CD (1998) Capture of an early fusion-active conformation of HIV-1 gp41. *Nat Struct Biol* 5: 276–279
14 He Y, Vassell R, Zaitseva M, Nguyen N, Yang Z, Weng Y, Weiss CD (2003) Peptides trap the human immunodeficiency virus type 1 envelope glycoprotein fusion intermediate at two sites. *J Virol* 77: 1666–1671
15 Wild C, Oas T, McDanal C, Bolognesi D, Matthews T (1992) A synthetic peptide inhibitor of human immunodeficiency virus replication: correlation between solution structure and viral inhibition. *Proc Natl Acad Sci USA* 89: 10537–10541
16 Jiang S, Lin K, Strick N, Neurath AR (1993) HIV-1 inhibition by a peptide. *Nature* 365: 113
17 Wild C, Greenwell T, Matthews T (1993) A synthetic peptide from HIV-1 gp41 is a potent inhibitor of virus-mediated cell-cell fusion. *AIDS Res Hum Retroviruses* 9: 1051–1053
18 Chan DC, Kim PS (1998) HIV entry and its inhibition. *Cell* 93: 681–684
19 Wild C, Greenwell T, Shugars D, Rimksy-Clark L, Matthews T (1995) The inhibitory activity of an HIV type 1 peptide correlates with its ability to interact with a leucine zipper structure. *AIDS Res Hum Retroviruses* 11: 323–325
20 Chen C-H, Matthews TJ, McDanal CB, Bolognesi DP, Greenberg ML (1995) A molecular clasp in the human immunodeficiency virus (HIV) type 1 TM protein determines the anti-HIV activity of gp41 derivatives: implication for viral fusion. *J Virol* 69: 3771–3777
21 Bewley CA, Louis JM, Ghirlando R, Clore GM (2002) Design of a novel peptide inhibitor of HIV fusion that disrupts the internal trimeric coiled-coil of gp41. *J Biol Chem*

22 Debnath AK (2006) Progress in identifying peptides and small-molecule inhibitors targeted to gp41 of HIV-1. *Expert Opin Investig Drugs* 15: 465–478
23 Kazmierski WM, Kenakin TP, Gudmundsson KS (2006) Peptide, peptidomimetic and small-molecule drug discovery targeting HIV-1 host-cell attachment and entry through gp120, gp41, CCR5 and CXCR4. *Chem Biol Drug Des* 67: 13–26
24 Wild CT, Shugars DC, Greenwell TK, McDanal CB, Matthews TJ (1994) Peptides corresponding to a predictive alpha-helical domain of human immunodeficiency virus type 1 are potent inhibitors of virus infection. *Proc Natl Acad Sci USA* 91: 9770–9774
25 Kilby JM, Hopkins S, Venetta TM, Di Massimo B, Cloud GA, Lee JY, Alldredge L, Hunter E, Lambert D, Bolognesi D et al. (1998) Potent suppression of HIV-1 replication in humans by T-20, a peptide inhibitor of gp41-mediated virus entry. *Nat Med* 4: 1302–1307
26 Chan DC, Chutkowski CT, Kim PS (1998) Evidence that a prominent cavity in the coiled coil of HIV type 1 gp41 is an attractive drug target. *Proc Natl Acad Sci USA* 95: 15613–15617
27 Rimsky LT, Shugars DC, Matthews TJ (1998) Determinants of human immunodeficiency virus type 1 resistance to gp41-derived inhibitory peptides. *J Virol* 72: 986–993
28 Wei X, Decker JM, Liu H, Zhang Z, Arani RB, Kilby JM, Saag MS, Wu X, Shaw GM, Kappes JC (2002) Emergence of resistant human immunodeficiency virus type 1 in patients receiving fusion inhibitor (T-20) monotherapy. *Antimicrob Agents Chemother* 46: 1896–1905
29 Koshiba T, Chan DC (2003) The prefusogenic intermediate of HIV-1 gp41 contains exposed C-peptide regions. *J Biol Chem* 278: 7573–7579
30 Root MJ, Hamer DH (2003) Targeting therapeutics to an exposed and conserved binding element of the HIV-1 fusion protein. *Proc Natl Acad Sci USA* 100: 5016–5021
31 Kilgore NR, Salzwedel K, Reddick M, Allaway GP, Wild CT (2003) Direct evidence that C-peptide inhibitors of human immunodeficiency virus type 1 entry bind to the gp41 N-helical domain in receptor-activated viral envelope. *J Virol* 77: 7669–7672
32 Trivedi VD, Cheng SF, Wu CW, Karthikeyan R, Chen CJ, Chang DK (2003) The LLSGIV stretch of the N-terminal region of HIV-1 gp41 is critical for binding to a model peptide, T20. *Protein Eng* 16: 311–317
33 Mink M, Mosier SM, Janumpalli S, Davison D, Jin L, Melby T, Sista P, Erickson J, Lambert D, Stanfield-Oakley SA et al. (2005) Impact of human immunodeficiency virus type 1 gp41 amino acid substitutions selected during enfuvirtide treatment on gp41 binding and antiviral potency of enfuvirtide *in vitro*. *J Virol* 79: 12447–12454
34 Nameki D, Kodama E, Ikeuchi M, Mabuchi N, Otaka A, Tamamura H, Ohno M, Fujii N, Matsuoka M (2005) Mutations conferring resistance to human immunodeficiency virus type 1 fusion inhibitors are restricted by gp41 and Rev-responsive element functions. *J Virol* 79: 764–770
35 Lawless MK, Barney S, Guthrie KI, Bucy TB, Petteway SR Jr, Merutka G (1996) HIV-1 membrane fusion mechanism: structural studies of the interactions between biologically-active peptides from gp41. *Biochemistry* 35: 13697–13708
36 Wexler-Cohen Y, Johnson BT, Puri A, Blumenthal R, Shai Y (2006) Structurally altered peptides reveal an important role for N-terminal heptad repeat binding and stability in the inhibitory action of HIV-1 peptide DP178. *J Biol Chem* 281: 9005–9010
37 Munoz-Barroso I, Durell S, Sakaguchi K, Appella E, Blumenthal R (1998) Dilation of the human immunodeficiency virus-1 envelope glycoprotein fusion pore revealed by the inhibitory action of a synthetic peptide from gp41. *J Cell Biol* 140: 315–323
38 Kliger Y, Gallo SA, Peisajovich SG, Munoz-Barroso I, Avkin S, Blumenthal R, Shai Y (2001) Mode of action of an antiviral peptide from HIV-1. Inhibition at a post-lipid mixing stage. *J Biol Chem* 276: 1391–1397
39 Liu S, Lu H, Niu J, Xu Y, Wu S, Jiang S (2005) Different from the HIV fusion inhibitor C34, the anti-HIV drug Fuzeon (T-20) inhibits HIV-1 entry by targeting multiple sites in gp41 and gp120. *J Biol Chem* 280: 11259–11273
40 Alam SM, Paleos CA, Liao HX, Scearce R, Robinson J, Haynes BF (2004) An inducible HIV type 1 gp41 HR-2 peptide-binding site on HIV type 1 envelope gp120. *AIDS Res Hum Retroviruses* 20: 836–845
41 Yuan W, Craig S, Si Z, Farzan M, Sodroski J (2004) CD4-induced T-20 binding to human immunodeficiency virus type 1 gp120 blocks interaction with the CXCR4 coreceptor. *J Virol* 78: 5448–5457
42 Dwyer JJ, Hasan A, Wilson KL, White JM, Matthews TJ, Delmedico MK (2003) The hydrophobic pocket contributes to the structural stability of the N-terminal coiled coil of HIV gp41 but is

not required for six-helix bundle formation. *Biochemistry* 42: 4945–4953
43 Derdeyn CA, Decker JM, Sfakianos JN, Zhang Z, O'Brien WA, Ratner L, Shaw GM, Hunter E (2001) Sensitivity of human immunodeficiency virus type 1 to fusion inhibitors targeted to the gp41 first heptad repeat involves distinct regions of gp41 and is consistently modulated by gp120 interactions with the coreceptor. *J Virol* 75: 8605–8614
44 Armand-Ugon M, Gutierrez A, Clotet B, Este JA (2003) HIV-1 resistance to the gp41-dependent fusion inhibitor C-34. *Antiviral Res* 59: 137–142
45 Lalezari JP, Bellos NC, Sathasivam K, Richmond GJ, Cohen CJ, Myers RA, Henry DH, Raskino C, Melby T, Murchison H et al. (2005) T-1249 retains potent antiretroviral activity in patients who had experienced virological failure while on an enfuvirtide-containing treatment regimen. *J Infect Dis* 191: 1155–1163
46 Menzo S, Castagna A, Monachetti A, Hasson H, Danise A, Carini E, Bagnarelli P, Lazzarin A, Clementi M (2004) Resistance and replicative capacity of HIV-1 strains selected *in vivo* by long-term enfuvirtide treatment. *New Microbiol* 27: 51–61
47 Reeves JD, Lee FH, Miamidian JL, Jabara CB, Juntilla MM, Doms RW (2005) Enfuvirtide resistance mutations: impact on human immunodeficiency virus envelope function, entry inhibitor sensitivity, and virus neutralization. *J Virol* 79: 4991–4999
48 Judice JK, Tom JY, Huang W, Wrin T, Vennari J, Petropoulos CJ, McDowell RS (1997) Inhibition of HIV type 1 infectivity by constrained alpha-helical peptides: implications for the viral fusion mechanism. *Proc Natl Acad Sci USA* 94: 13426–13430
49 Jin BS, Ryu JR, Ahn K, Yu YG (2000) Design of a peptide inhibitor that blocks the cell fusion mediated by glycoprotein 41 of human immunodeficiency virus type 1. *AIDS Res Hum Retroviruses* 16: 1797–1804
50 Sia SK, Carr PA, Cochran AG, Malashkevich VN, Kim PS (2002) Short constrained peptides that inhibit HIV-1 entry. *Proc Natl Acad Sci USA* 99: 14664–14669
51 Otaka A, Nakamura M, Nameki D, Kodama E, Uchiyama S, Nakamura S, Nakano H, Tamamura H, Kobayashi Y, Matsuoka M, Fujii N (2002) Remodeling of gp41-C34 peptide leads to highly effective inhibitors of the fusion of HIV-1 with target cells. *Angew Chem Int Ed Engl* 41: 2937–2940
52 Wang S, York J, Shu W, Stoller MO, Nunberg JH, Lu M (2002) Interhelical interactions in the gp41 core: implications for activation of HIV-1 membrane fusion. *Biochemistry* 41: 7283–7292
53 Eckert DM, Malashkevich VN, Hong LH, Carr PA, Kim PS (1999) Inhibiting HIV-1 entry: discovery of D-peptide inhibitors that target the gp41 coiled-coil pocket. *Cell* 99: 103–115
54 Ferrer M, Kapoor TM, Strassmaier T, Weissenhorn W, Skehel JJ, Oprian D, Schreiber SL, Wiley DC, Harrison SC (1999) Selection of gp41-mediated HIV-1 cell entry inhibitors from biased combinatorial libraries of non-natural binding elements. *Nat Struct Biol* 6: 953–960
55 Debnath AK, Radigan L, Jiang S (1999) Structure-based identification of small molecule antiviral compounds targeted to the gp41 core structure of the human immunodeficiency virus type 1. *J Med Chem* 42: 3203–3209
56 Jiang S, Lin K, Zhang L, Debnath AK (1999) A screening assay for antiviral compounds targeted to the HIV-1 gp41 core structure using a conformation-specific monoclonal antibody. *J Virol Methods* 80: 85–96
57 Jiang S, Debnath AK (2000) A salt bridge between an N-terminal coiled coil of gp41 and an antiviral agent targeted to the gp41 core is important for anti-HIV-1 activity. *Biochem Biophys Res Commun* 270: 153–157
58 Armand-Ugon M, Clotet-Codina I, Tintori C, Manetti F, Clotet B, Botta M, Este JA (2005) The anti-HIV activity of ADS-J1 targets the HIV-1 gp120. *Virology* 343: 141–149
59 Zhao Q, Ernst JT, Hamilton AD, Debnath AK, Jiang S (2002) XTT formazan widely used to detect cell viability inhibits HIV type 1 infection *in vitro* by targeting gp41. *AIDS Res Hum Retroviruses* 18: 989–997
60 Jiang S, Lu H, Liu S, Zhao Q, He Y, Debnath AK (2004) N-Substituted pyrrole derivatives as novel human immunodeficiency virus type 1 entry inhibitors that interfere with the gp41 six-helix bundle formation and block virus fusion. *Antimicrob Agents Chemother* 48: 4349–4359
61 Labrosse B, Pleskoff O, Sol N, Jones C, Henin Y, Alizon M (1997) Resistance to a drug blocking human immunodeficiency virus type 1 entry (RPR103611) is conferred by mutations in gp41. *J Virol* 71: 8230–8236
62 Jin BS, Lee WK, Ahn K, Lee MK, Yu YG (2005) High-throughput screening method of inhibitors that block the interaction between 2 helical regions of HIV-1 gp41. *J Biomol Screen* 10: 13–19

63 Lipinski CA (2000) Drug-like properties and the causes of poor solubility and poor permeability. *J Pharmacol Toxicol Methods* 44: 235–249

64 Eckert DM, Kim PS (2001) Design of potent inhibitors of HIV-1 entry from the gp41 N-peptide region. *Proc Natl Acad Sci USA* 98: 11187–11192

65 Lu M, Kim PS (1997) A trimeric structural subdomain of the HIV-1 transmembrane glycoprotein. *J Biomol Struct Dyn* 15: 465–471

66 Bianchi E, Finotto M, Ingallinella P, Hrin R, Carella AV, Hou XS, Schleif WA, Miller MD, Geleziunas R, Pessi A (2005) Covalent stabilization of coiled coils of the HIV gp41 N region yields extremely potent and broad inhibitors of viral infection. *Proc Natl Acad Sci USA* 102: 12903–12908

67 Root MJ, Kay MS, Kim PS (2001) Protein design of an HIV-1 entry inhibitor. *Science* 291: 884–888

68 Louis JM, Bewley CA, Clore GM (2001) Design and properties of N(CCG)-gp41, a chimeric gp41 molecule with nanomolar HIV fusion inhibitory activity. *J Biol Chem* 276: 29485–29489

69 Greenberg M, Cammack N, Salgo M, Smiley L (2004) HIV fusion and its inhibition in antiretroviral therapy. *Rev Med Virol* 14: 321–337

70 Greenberg ML, Cammack N (2004) Resistance to enfuvirtide, the first HIV fusion inhibitor. *J Antimicrob Chemother* 54: 333–340

71 Sista PR, Melby T, Davison D, Jin L, Mosier S, Mink M, Nelson EL, De Masi R, Cammack N et al. (2004) Characterization of determinants of genotypic and phenotypic resistance to enfuvirtide in baseline and on-treatment HIV-1 isolates. *AIDS* 18: 1787–1794

72 Perez-Alvarez L, Carmona R, Ocampo A, Asorey A, Miralles C, Perez de Castro S, Pinilla M, Contreras G, Taboada JA, Najera (2006) Long-term monitoring of genotypic and phenotypic resistance to T20 in treated patients infected with HIV-1. *J Med Virol* 78: 141–147

73 Baldwin CE, Sanders RW, Deng Y, Jurriaans S, Lange JM, Lu M, Berkhout B (2004) Emergence of a drug-dependent human immunodeficiency virus type 1 variant during therapy with the T20 fusion inhibitor. *J Virol* 78: 12428–12437

74 Xu L, Pozniak A, Wildfire A, Stanfield-Oakley SA, Mosier SM, Ratcliffe D, Workman J, Joall A, Myers R, Smit T E et al. (2005) Emergence and evolution of enfuvirtide resistance following long-term therapy involves heptad repeat 2 mutations within gp41. *Antimicrob Agents Chemother* 49: 1113–1119

75 Jenwitheesuk E, Samudrala R (2005) Heptad-repeat-2 mutations enhance the stability of the enfuvirtide-resistant HIV-1 gp41 hairpin structure. *Antiviral Ther* 10: 893–900

76 Poveda E, Rodes B, Lebel-Binay S, Faudon JL, Jimenez V, Soriano V (2005) Dynamics of enfuvirtide resistance in HIV-infected patients during and after long-term enfuvirtide salvage therapy. *J Clin Virol* 34: 295–301

77 Lu J, Sista P, Giguel F, Greenberg M, Kuritzkes DR (2004) Relative replicative fitness of human immunodeficiency virus type 1 mutants resistant to Enfuvirtide (T-20). *J Virol* 78: 4628–4637

78 Lohrengel S, Hermann F, Hagmann I, Oberwinkler H, Scrivano L, Hoffmann C, von Laer D, Dittmar MT (2005) Determinants of human immunodeficiency virus type 1 resistance to membrane-anchored gp41-derived peptides. *J Virol* 79: 10237–10246

79 Neumann T, Hagmann I, Lohrengel S, Heil ML, Derdeyn CA, Krausslich HG, Dittmar MT (2005) T20-insensitive HIV-1 from naive patients exhibits high viral fitness in a novel dual-color competition assay on primary cells. *Virology* 333: 251–262

80 Reeves JD, Gallo SA, Ahmad N, Miamidian JL, Harvey PE, Sharron M, Pohlmann, Sfakianos JN, Derdeyn CA, Blumenthal R et al. (2002) Sensitivity of HIV-1 to entry inhibitors correlates with envelope/coreceptor affinity, receptor density, and fusion kinetics. *Proc Natl Acad Sci USA* 99: 16249–16254

81 Beausejour Y, Tremblay MJ (2004) Susceptibility of HIV type 1 to the fusion inhibitor T-20 is reduced on insertion of host intercellular adhesion molecule 1 in the virus membrane. *J Infect Dis* 190: 894–902

82 Reeves JD, Miamidian JL, Biscone MJ, Lee FH, Ahmad N, Pierson TC, Doms RW (2004) Impact of mutations in the coreceptor binding site on human immunodeficiency virus type 1 fusion, infection, and entry inhibitor sensitivity. *J Virol* 78: 5476–5485

83 Miller MD, Hazuda DJ (2004) HIV resistance to the fusion inhibitor enfuvirtide: mechanisms and clinical implications. *Drug Resist Updat* 7: 89–95

84 Desmezieres E, Gupta N, Vassell R, He Y, Peden K, Sirota L, Yang Z, Wingfield P, Weiss CD (2005) HIV gp41 escape mutants: Cross-resistance to peptide inhibitors of HIV fusion and altered

receptor activation of gp120. *J Virol* 79: 4774–4781
85 Steger HK, Kaur J, Reeves J, Doms RW, Root MJ (2005) *Keystone Symposium HIV Pathogenesis*, abstract no. 346
86 Hanna SL, Yang C, Owen SM, Lal RB (2002) Variability of critical epitopes within HIV-1 heptad repeat domains for selected entry inhibitors in HIV-infected populations worldwide. *AIDS* 16: 1603–1608
87 Labrosse B, Labernardiere J-L, Dam E, Trouplin V, Skrabal K, Clavel F, Mammano F (2003) Baseline susceptibility of primary human immunodeficiency virus type 1 to entry inhibitors. *J Virol* 77: 1610–1613
88 Chinnadurai R, Munch J, Kirchhoff F (2005) Effect of naturally-occurring gp41 HR1 variations on susceptibility of HIV-1 to fusion inhibitors. *AIDS* 19: 1401–1405
89 Sanders RW, Korber B, Lu M, Berkhout B, Moore JP (2002) Mutational analysis and natural variability of the gp41 ectodomain. In: Kuiken C, Foley B, Freed E, Hahn B, Marx P, McCutchan F, Mellors J, Wolinsky S, Korber B (eds): HIV sequence compendium 2002. Los Alamos National Laboratory Theoretical Biology and Biophysics Group, Los Alamos, 43–68
90 Derdeyn CA, Decker JM, Sfakianos JN, Wu X, O'Brien WA, Ratner L, Kappes JC, Shaw GM, Hunter E (2000) Sensitivity of human immunodeficiency virus type 1 to the fusion inhibitor T-20 is modulated by coreceptor specificity defined by the V3 loop of gp120. *J Virol* 74: 8358–8367
91 Heil ML, Decker JM, Sfakianos JN, Shaw GM, Hunter E, Derdeyn CA (2004) Determinants of human immunodeficiency virus type 1 baseline susceptibility to the fusion inhibitors enfuvirtide and T-649 reside outside the peptide interaction site. *J Virol* 78: 7582–7589
92 Barretina J, Blanco J, Bonjoch A, Llano A, Clotet B, Este JA (2004) Immunological and virological study of enfuvirtide-treated HIV-positive patients. *AIDS* 18: 1673–1682
93 Eron JJ, Gulick RM, Bartlett JA, Merigan T, Arduino R, Kilby JM, Yangco B, Diers A, Drobnes C, DeMasi R et al. (2004) Short-term safety and antiretroviral activity of T-1249, a second-generation fusion inhibitor of HIV. *J Infect Dis* 189: 1075–1083
94 Ji H, Shu W, Burling FT, Jiang S, Lu M (1999) Inhibition of human immunodeficiency virus type 1 infectivity by the gp41 core: role of a conserved hydrophobic cavity in membrane fusion. *J Virol* 73: 8578–8586

Entry Inhibitors in HIV Therapy
Edited by Jacqueline D. Reeves and Cynthia A. Derdeyn

HIV-1 entry inhibitors as microbicides

Clyde E. Hart and Tammy Evans-Strickfaden

Division of HIV/AIDS Prevention, Center for HIV, Viral Hepatitis, STD & TB Prevention, Centers for Disease Control and Prevention, Atlanta, GA 30333, USA

Introduction

Sexual transmission of HIV-1 is the major route for infection of both men and women [1]. During the early phase of the HIV pandemic in the western hemisphere and industrialized world, sexual transmission of HIV was regarded as a public health concern that mostly impacted men who have sex with men (MSM) and commercial sex workers and their male clients [2]. In the developing world, however, heterosexual transmission of HIV between sex partners both outside and within the context of the commercial sex industry was identified early on as a major driving force for the catastrophic rate of infection now reported in many regions of the world [2, 3].

The recognition that sexual transmission of HIV is the major mode for infection globally prompted international agencies and national and local governments to establish behavioral intervention programs that promoted male condom use and sexual abstinence outside of a monogamous relationship. These types of intervention programs have reduced HIV infection rates among some high-risk groups when those programs have had sustained local community support. In particular, aggressive promotion of male condom use in the commercial sex trade within some countries and abstinence, be faithful, or use a condom (ABC) strategies for sexually active persons outside the sex trade in other countries have lowered HIV-1 transmission rates [4–6]. However, in many developing countries, and in particular those with the highest HIV infection rates, the promotion of behavioral intervention programs, including male condom use among sexually active couples outside the commercial sex trade has found limited success. The gender inequality, societal, and interpersonal intimacy issues that surround the refusal of male condom use by sexual partners at high risk for HIV transmission is a complex problem of grave public health importance [7]. Because of this, and in the absence of an available preventative vaccine in the near future, the development of additional behavioral and biomedical strategies to reduce HIV sexual transmission have taken on added importance.

Even during the early stages of the HIV epidemic when researchers and policy makers predicted a relatively swift development of an effective preventative vaccine, calls came forth for the development of vaginally applied microbicides that could be used by a woman to protect her and her sexual partner from sexually transmitted infections, including HIV [8, 9]. The first compound tested in clinical trials was the detergent-based N-9 spermacide, which was found to increase HIV-1 transmission [10]. In response, the newly emerging field of HIV-1 microbicide research brought forward alternative candidate microbicides that, for the most part, were extracts of naturally occurring compounds. Many of these candidate microbicides are polyanionic compounds whose anti-infective activity in cell culture is reported to be the inhibition of cell-free HIV-1 entry into target cells [11]. However, not long after these, synthetic polyanionic compounds with similar HIV-1 anti-entry/fusion activities were introduced as candidate microbicides. In fact, at present, only polyanionic compounds (one natural, one synthetic) are undergoing Phase III efficacy testing in human clinical trials [12].

Only recently have microbicide intervention strategies expanded to the extent that are warranted within the HIV research community. Some of the more recently proposed topical microbicide candidates have previously been investigated as orally administered experimental therapeutic drugs for infected persons. These include HIV-1 replication inhibitors [13, 14], chemokine receptor blockers [15–17], and HIV-1 envelope protein ligands [18, 19]. Other candidate microbicides have been specifically developed for use as a topical microbicide; these include vaginal pH buffering compounds [20], genetically engineered H_2O_2-producing lactobacilli [21–23], detergents [24], anti-HIV-1 antibodies [25], carbohydrate-binding lectins [26], herbal extracts [27], and polyanionic polymers [11]. The anti-infective activity reported for many of these candidate products is inhibition of attachment, entry, or fusion of cell-free HIV-1 with target cells. Depending on the compound, this activity can be an HIV-1-specific or nonspecific mechanism and can target cellular or viral epitopes. In light of the significant role for binding/entry/fusion inhibitors as candidate microbicides, this chapter reviews the progress and challenges for their use as topical agents to block sexual transmission of HIV-1.

Female genital tract targets for blocking HIV-1 entry/fusion

In the female genital tract there are different tissue sites and multiple cell types that have been proposed to be necessary for establishing an HIV-1 infection in women through sexual transmission. Human cervical and vaginal mucosal secretions contain HIV-1 target cells such as $CD4^+$ T lymphocytes. In addition, these lymphocytes are found within cervical and vaginal epithelia [28]. Also within these epithelia are the $CD4^+CCR5^+$ Langerhans cells that are capable of transporting cell-free HIV-1 from the genital tract lumen to sub-epithelial HIV-1 target cells, including $CD4^+$ T lymphocytes, macrophages, and sub-epithelial dendritic cells (Fig. 1) [28, 29].

Genital tract infection studies using female non-human primates (NHPs) report that $CD4^+$ cells in the cervical and vaginal epithelium are the first cell

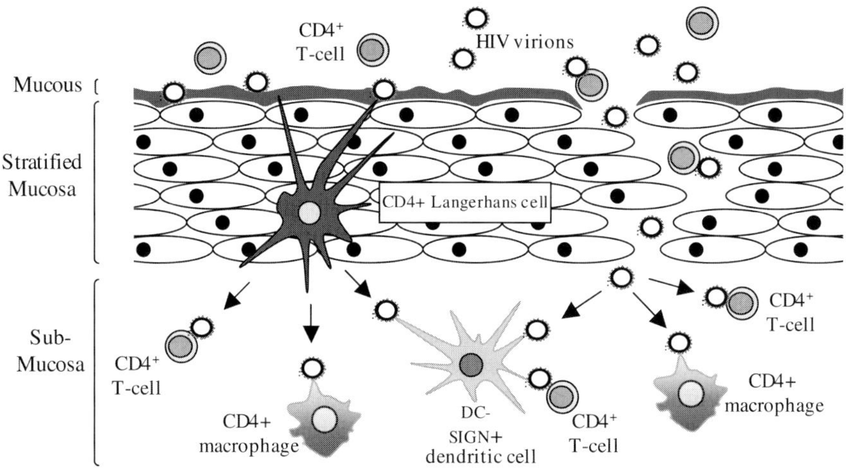

Figure 1. Possible pathways for sexual transmission of HIV-1 in the female genital tract. The human vaginal and ectocervical mucosa consists of a multilayer (usually 30 to 45 cell layers) of nonkeratinized squamous epithelial cells covered by a mucous that most likely originates from cervical mucous glands. The endocervical mucosa is a single layer of mucous-containing, high columnar cells (not shown). When HIV-1 is deposited in the vagina, it may bind to antigen-presenting CD4$^+$, CCR5$^+$ Langerhans cells that have cellular processes that extend to the vaginal lumen. The Langerhan cells are then capable of transferring the virus to CD4$^+$ T cells and macrohpages, and DC-SIGN$^+$ dendritic cells in the submucosal lamina propria. Cell-free HIV-1 may also directly infect CD4$^+$ T cells in the vaginal lumen that have migrated up through the stratified epithelium or that have passed through mucosal lesions. In addition, mucosal lesions may provide a ready accessible pathway for HIV-1 or HIV-1-infected cells to pass easily into the submucosa that contains numerous CD4$^+$ and DC-SIGN$^+$ target cells. From the sub-mucosa, infected cells can migrate to nearby lymphoid tissues where virus replication would be expanded and systemic dissemination of HIV-1 can occur.

populations infected after non-traumatic vaginal inoculation with cell-free simian immunodeficiency virus (SIV), the NHP equivalent of HIV-1. The available limited time course data suggest that cell-free SIV can enter the genital epithelia and infect cells within 30–60 minutes after vaginal virus exposure. *In situ* immunocytochemical staining of fixed cervical and vaginal tissues as early as 18 h post-SIV vaginal inoculation show that CD4$^+$ lymphocytes and intra-epithelial CD4$^+$ Langerhans cells are the first to be positive for SIV RNA. Later time points show SIV RNA-positive CD4$^+$ lymphocytes and macrophages, and dendritic cells in the sub-epithelial lamina propria of cervical-vaginal mucosa [30].

The above SIV cervical-vaginal infection studies in NHPs attempt to predict how a male-to-female sexual infection occurs when the genital tract epithelium is intact. However, the likelihood of normally occurring mucosal lesions in the female vagina and those produced during coitus or from local inflammation due to ulcerative non-HIV sexually transmitted infections (STIs) would provide a more direct route for HIV-1 to breach the genital tract epithelium and access the numerous CD4$^+$ lymphocyte and macrophage virus targets in the lamina propria. In fact, data from studies where the human female genital tract was sampled using a buffered saline lavage show that blood and leukocytes are present on the cervical and vaginal mucosal surfaces of many women at intermenses time points [31, 32]. These observations suggest that minor mucosal lesions probably do provide a significant pathway for direct sub-epithelial access of HIV-1 to target cells in the genital tract of women even in the absence of ulcerative STIs. Thus, both genital mucosal lesions and the transfer of virus across an intact epithelium by Langerhans cells should be considered as probably pathways for the sexual transmission of HIV-1 to women.

Fortunately, most of the HIV-1 entry/fusion inhibitors that are currently under consideration as microbicides should have anti-infective activities that are effective in the presence or absence of an intact genital epithelium (Fig. 2). The following is a description of some of the most promising candidate entry/fusion inhibitors, their proposed mechanisms of action, and their likely advantages and disadvantages for success at blocking HIV-1 sexual transmission.

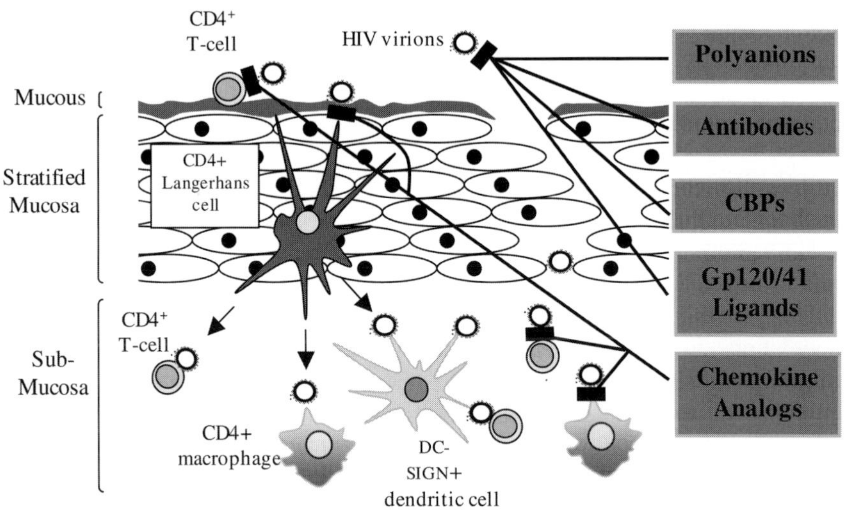

Figure 2. Predicted sites of activity for HIV-1 entry inhibitors as vaginal microbicides. Many of the currently know entry inhibitors being considered as a microbicide function by binding to one of the HIV-1 envelope proteins (gp120 or gp41), or to the cell-surface chemokine receptor (CCR5) that is also the virus co-receptor associated with most cases of sexual transmission. Most polyanions, human monoclonal antibodies, carbohydrate binding proteins (CBPs), and gp120/41 ligands are believed to inactivate HIV-1 by binding to its surface proteins and blocking the interaction with the CD4^{+} receptor and, or CCR5 co-receptor. These candidate products would be predicted to work best against cell-free HIV-1 in the vaginal lumen but they may be able to diffuse through mucosal lesions and have some limited activity during virus binding to targets cells. Conversely, the chemokine analogs currently under consideration would work by binding to, and either down regulating cell surface expression of the CCR5 co-receptor or inducing it to form an unusable configuration. The chemokine analogs would have to be available to cells both in the vaginal lumen and submucosa if cell-free virus can infect target cells at both sites.

Polyanionic polymers

Polyanionic polymers came to prominence as some of the first generation of candidate microbicides. This group of candidate products consists of compounds derived from naturally occurring substances and synthetic polymers. Presently, they compose the largest group of lead investigational products that are currently undergoing clinical testing in humans (Tab. 1). For most of these compounds, their proposed mechanism of action is to bind the gp120 envelope protein on the virion surface and prevent binding to the surface of target cells. *In vitro* testing of sulfated polyanionic polymers indicates that their negatively charged side groups have a significant affinity for the positively charged V3 loop and to a lesser degree the CD4-induced epitope, both of which are contained within the viral gp120 envelope protein [33]. The V3 loop and the CD4-induced epitope are recognized as the principal regions

Table 1. HIV-1 entry inhibitors under investigation as vaginal microbicides

Product	Proposed activity/target	Testing status
Antibodies		
b 12	Virus gp 120	Animal (primate)
2G12	Virus gp120 carbohydrates	Animal (primate)
2F5	Virus gp 41	Animal (primate)
Chemokine analogs		
CMPD 167	CCR5 antagonist	Animal (primate)
PSC RANTES	Down-regulates CCR5	Animal (primate)
gp-120/gp-41 ligands		
BMS 378806	Virus gp120	Animal (primate)
Pro542	Virus gp120 (CD4 binding site)	Cervical explants
C52L	Virus gp41	Animal (primate)
Carbohydrate-binding proteins		
Cyanovirin-N	gp120 carbohydrates	Animal (mouse)
Soluble DC-SIGN	gp120 carbohydrates	Unknown
Polyanions		
Carraguard	V3 loop of gp120	Phase III
Cellulose sulfate	"	Phase III*
Cellulose acetate 1,2-benzenedicarboxylate	"	Animal (primate)
Dextrin-2-sulfate	"	Phase I*
Polystyrene sulfonate	"	Phase I*
Pro2000	"	Phase III
SPL7013	"	Phase I
SAMMA	viral envelope	*In vitro*

* Clinical trial halted

within the viral gp120 that bind to HIV-1 co-receptors on target cells. In addition to their projected potential for blocking sexual transmission of HIV-1, many of the polyanionic polymers have excellent laboratory profiles as potent inhibitors of other STIs such as herpes simplex virus, human papilloma virus, gonorrhea, and Chlamydia. The anticipated performance of these candidate products to block sexual transmission of both HIV-1 and other STIs associated with increased transmission of HIV-1 was a major factor in their advancement to clinical trials.

The three polyanionic polymers that are being, or have been, tested in Phase III clinical trials are cellulose sulfate (Ushercell^{TM}), PC515 (Carraguard^{TM}), and polynaphthalene sulfonate (Pro 2000). Cellulose sulfate (CS) is a sulfated carboxymethylcellulose polymer (1900 kDa) extracted from cotton. More often recognized by its trade name, Carraguard^{TM}, PC515 is also a sulfated long-change carbohydrate (500–5000 kDa) composed of lambda and kappa carrageenans extracted from red seaweed. Carrageenan is widely used as an emulsifying agent in food and cosmetics, and is listed as a GRAS (generally regarded as safe) agent by the USA FDA. Pro 2000 is the first synthetic polyanionic polymer, a polysulfonic acid-formaldehyde copolymer polymer (5 kDa) to be tested in a Phase III trial. Before their use in Phase III clinical trials, these products were shown to have good safety profiles in animal testing and Phase I/II clinical safety trials. However, their success as anti-infective agents has been more difficult to predict due to our limited understanding of which, if any, of the *in vitro* experiments and animal transmission studies used to investigate these products are indicative of their effectiveness when used by women.

In vitro testing of these products has used a variety of cellular infection assays. Overall, the outcome of these experiments has been determined by measuring the difference in cell-free HIV-1 production after cell cultures are inoculated with virus in the absence or presence of a candidate product. CS, Carraguard^{TM}, and Pro 2000 have all been reported to significantly reduce HIV-1 production from cells exposed to virus in their presence. However, each of these candidate products has shown different *in vitro* capacities for blocking infection from CXCR4(X4)- and CCR5(R5)-using HIV-1. Since most of the documented HIV-1 infections acquired from sexual transmission result from R5-using viruses [34], the ability of a candidate microbicide to act against those virus types *in vitro* has been regarded as one of the major criteria for predicting its effectiveness in clinical trials.

For CS, the reported IC_{50} values are similar against the laboratory strains of X4-using HIV-1_{IIIB} (IC_{50} = 10 μg/ml) and R5-using HIV-1_{BaL} (IC_{50} = 8.8 μg/ml) when tested in primary $CD4^+$ T lymphocytes and macrophages, respectively [35]. A previous study [36] also reported equivalent *in vitro* IC_{50} values for CS activities against HIV-1_{IIIB} (32 μg/ml), HIV-1_{BaL} (78 μg/ml) and another R5-using isolate (ADA, 1.3 μg/ml), suggesting that this product has the predicted capacity to be effective against sexually transmitted R5-using HIV-1. Surprisingly, CS was even more effective *in vitro* against R5-

Human monoclonal antibodies

Three human monoclonal antibodies (mAbs), b12, 2G12, and 2F5, have received the most attention as "proof-of-concept" reagents for blocking sexual transmission of HIV-1 in animal studies. The mAb b12 binds to discontinuous epitopes in the HIV-1 gp120 protein that overlap the CD4-binding site [57, 58]; mAb 2G12 recognizes a unique conformational glycan epitope consisting of the terminal mannose residues on three *N*-linked carbohydrates in gp120 [59, 60]; and mAb 2F5 binds epitopes on the external surface of gp41 [61, 62]. Additional information about these antibodies is provided in the chapter by Morris et al.

Of these three mAb, only b12 has been examined as a vaginally applied anti-infective in the Rhesus macaque model. When a high concentration of mAb b12 in either saline or a hydroxymethyl cellulose gel was applied intravaginally, it blocked infection of R5 SHIV intravaginal inoculations in 9 of 12 macaques [63]. The mAbs 2G12 and 2F5 have also been tested in the macaque model but using intravenous infusion instead of intravaginal application [64]. However, mAb 2G12 infused alone or in combination with 2F5 provided only partial protection (8/14 animals) against X4 SHIV vaginal inoculations [64]. Previous *in vitro* data show that these mAbs have high titer neutralization activities against most HIV-1 subtype B isolates tested, but only limited activity against non-subtype B viruses [43, 65–69]. Although these data suggest that mAbs have the potential to be developed as effective microbicides, a new generation of these candidate products that have efficacy against non-subtype B viruses will be needed. Also, development of new cost-effective methods for their large-scale production, such as production as plantibodies [70], will be a practical necessity.

Carbohydrate-binding proteins

Oligosaccharide carbohydrates are predicted to be more than half of the mass of the gp120 and gp41 envelope glycoproteins on the surface of HIV-1 [71]. Many of these oligosaccharides are large poly-mannose containing sugars [72, 73] that are believed to form a virus "glycan shield" for providing protection against some host defense mechanisms [71]. The relative continuity of this mannose-rich glycan shield across the different HIV-1 subtypes has made it a tempting target for developing candidate microbicides. In addition to the human mAb 2G12 described above that recognizes a site-specific epitope of mannose residues on the gp120 glycoprotein, a number of carbohydrate-binding proteins (CBPs) derived from prokaryotes, sea corals, algae, plants, and vertebrates with anti-HIV-1 activities have been reported (reviewed in [74]) and many have shown anti-infective protection *in vitro* against multiple HIV-1 subtypes and HIV-2 [75].

The 11-kDa Cyanovirin-N (CV-N) protein, originally extracted from blue green algae, has been the most intensely studied anti-HIV-1 CBP. CV-N has excellent anti-infective activity that is reported to occur through its high-affinity binding to high mannose oligosaccharides on the gp120 protein of HIV-1 [76]. CV-N's binding to these mannose glycans blocks gp120 binding to cell-associated CD4 and possibly to the CCR5 co-receptor *in vitro* [76]. CV-N is effective at blocking R5 HIV-1 infection of human cervical explants in a dose- and time-dependent manner; it is more efficacious when applied at a high dose (1–0.2 μM) before virus exposure [77]. In addition, it has virucidal activity when used to pretreat a virus inoculum and is subsequently removed before the inoculum is added to an explant [77]. When CV-N (0.5–2%) in a hydroxyethyl cellulose gel was applied intravaginally in cynomologus macaques, it blocked infection of intravaginal inoculations of high titer X4 SHIV in 15 of 18 animals [77]. The excellent anti-infective profile of CV-N protein described above and its broad spectrum activity across HIV-1 subtypes and HIV-2 [78] have led to it being engineered into a human vaginal strain of *Lactobacillus* [22]. This CV-N-expressing *Lactobacillus* was selected for its capacity as an efficient protein expression vehicle and because it is a naturally occurring vaginal species that is associated with reduced urogenital infections in women [79]. The CV-N-expressing strain is capable of colonizing the vagina and producing active CV-N protein when administered intravaginally in mice [22]. This development of a CV-N-expressing *Lactobacillus* is a significant step in advancing microbicide products that can possibly be maintained for extended periods in the female genital tract.

In addition to the potential for CBPs to act directly as entry inhibitors against HIV-1 infection of CD4 target cells in the genital mucosa, they may also have the capacity to prevent infection in these cells via transfer of HIV from CD4-/DC-SIGN$^+$ dendritic cells. As briefly mentioned above, submucosal dendritic cells are capable of binding HIV on their cell surface, and the bound virus can be presented to, and *trans* infect, CD4 lymphocytes [80]. The attachment of HIV to dendritic cells is reported to be dependent on the binding of virus polymannose carbohydrates in the gp120 protein to the DC-SIGN (C-type lectin) molecule expressed on the dendritic cell surface [81, 82]. These plant-derived CBPs and CV-N effectively block virus binding to cell surface DC-SIGN and the subsequent *trans* infection of CD4 lymphoctyes *in vitro* [83]. Thus, CBP microbicide products could have the advantage of a dual mechanism for blocking HIV-1 infection in the genital mucosa of women. However, the safety of repeated applications, or long-term mucosal exposure of a CBP protein could elicit local immune responses in the genital tract of women. For example, the pronounced mitogenic activity and induction of activation markers on human peripheral blood leukocytes *in vitro* [78] by CV-N indicates that any CBP candidate microbicide would require stringent monitoring during *in vivo* safety studies.

The potential safety concern of using non-human CBPs for microbicides suggests that the development of similar products of human origin should be considered. Of particular interest is the human DC-SIGNR molecule that has R5 and X4 HIV-1 gp120-binding affinities across subtypes that are almost identical to those of its close homolog DC-SIGN [84]. Multiple mRNA isoforms of DC-SIGNR have recently been identified in human vaginal tissues and many of these were predicted to code for soluble and secreted proteins [85]. In addition, different DC-SIGNR isoforms have been correlated with possible resistance to HIV-1 infection [86]. The identification of soluble DC-SIGNR isoforms with effective anti-HIV-1 activities *in vitro* and their expression in *Lactobacillus* could provide a safer alternative to the use of non-human CBPs as microbicides.

Specific ligands of HIV-1 gp120 and gp41 envelope proteins

These small molecule and fusion protein drugs are part of a second generation of candidate microbicides that were developed to have anti-HIV-1-specific activities. In most cases, they were initially designed and tested as therapeutic drugs for treating infected persons before they were examined as topical agents for blocking sexual transmission of HIV-1. Although numerous gp120 and gp41 ligands have been tested for anti-HIV-1 therapeutic activity, those described below have been evaluated specifically for their effectiveness as topical microbicide agents.

Two of these ligands, BMS 378806 and Pro542, specifically bind to the CD4 binding site of the HIV-1 gp120 protein and prevent virus entry [87, 88]. BMS 378806 is a small molecule that works allosterically to prevent gp120 conformational changes required for its binding to CD4 [89] and Pro542 is a polyvalent CD4-IgG2 fusion protein that has a much longer half life *in vivo* than soluble CD4 alone [87]. Alternatively, the C52L peptide is a gp41-binding ligand that was derived as a sequence modification of the FDA-approved T20/enfuvirtide peptide [90]. These ligands block formation of virus gp41 six-helix bundles that are required for fusion between the virus and cell membranes [19]. BMS 378806 and C52L have anti-infective activities across multiple HIV-1 subtypes, but BMS 378806 activity is much greater against subtype B virus [91, 92]. Pro542 anti-infective activity in human cervical explants indicates that it can block both localized infection within the tissue and *trans* infection via dissemination of dendritic cell-bound infectious virus away from the tissue explant [93]. Pro542 binding to, and neutralization of, cell-free virus that can still bind to migratory $CD4^-$/DC-$SIGN^+$ dendritic cells is proposed to underlie its ability to block the *trans* infection of distant $CD4^+$ target cells [93]. Both BMS 378806 and C52L have been tested alone or in combination in the macaque vaginal infection model. A high concentration of C52L (1.5 mM) or BMS 378806 (5.5 mM) in a hydroxmethyl cellulose gel provided partial protection from vaginal inoculations of an R5 SHIV in three of five and six of eight macaques, respectively [94]. BMS 378806 and C52L used in combination increased their effectiveness such that all the macaques tested were protected from R5 SHIV infection (six of six protected, [94]). In vaginal safety studies using macaques, neither drug induced genital irritation or inflammation after multiple applications [94]. The original development of these anti-HIV-1 drugs as therapeutic agents was done under the constraints of needing to balance their potency and oral bioavailability. It is hoped that newer and more effective derivates of this class of ligand will be tested as mucosally applied products and thus consideration of their oral bioavailability will be irrelevant. In fact, the lack of systemic absorption of topically applied candidate microbicides is currently regarded a positive safety criterion.

Chemokine analogs

Chemokine analogs are another class of anti-HIV-1 drugs that were developed as therapeutic agents. The discovery that certain native chemokines have anti-infective activity in cell culture led to the identification of two of their cell surface receptors, CCR5 and CXCR4, as the major co-receptors for HIV-1 binding and entry [95–99]. The testing of these analogs as candidate microbicides have focused on those that block CCR5 binding since R5 viruses are believed to cause the vast majority HIV-1 sexual transmissions [100, 101].

Only recently has a chemokine analog been investigated alone or in combination with other compounds for use as a vaginal microbicide. The PSC RANTES analog is a modified synthetic form of the native RANTES ligand that binds to the CCR5 receptor on the cell surface and induces its internalization [102]. This ligand-induced receptor internalization greatly reduces the level of cell surface CCR5 and is thought to be the major mechanism for chemokine anti-HIV-1 activity [103]. Compared to native RANTES, the synthetic PSC RANTES analog greatly enhances receptor internalization and has several-fold more anti-HIV-1 activity *in vitro* [102]. *In vivo*, vaginal applications of the highest concentration of PSC RANTES (1 mM) used prevented infection from vaginal inoculation of an R5 SHIV in five of five macaques [104]. Another CCR5-binding ligand tested as a candidate microbicide is the small molecule CMPD 167. It is proposed to function as an allosteric modulator by binding to CCR5 and locking it into a conformation that is unusable as an HIV-1 co-receptor [105]. As seen with PSC RANTES, a high concentration of CMPD 167 (5 mM) alone or in combination with C52L or C52L plus BMS 378806 in hydroxmethyl cellulose gel protected seven of seven macaques from infection with an intravaginal inoculation of R5 SHIV [94].

These results suggest that chemokine analogs that block or down-regulate the HIV-1 CCR5 coreceptor are viable microbicide candidates that deserve continued evaluation. Although the majority of HIV-1 sexual transmissions are believed to result from infection with R5 virus, the development of a microbicide product that does not block X4 virus infections or virus binding to DC-SIGN^{+} dendritic cells would have to be carefully considered before use in human clinical trials. Also, as with other drug-based products proposed for mucosal application, the potential for systemic absorption and toxicity will have to be taken in to account, in addition to formulating a gel or delivery system that will rapidly and evenly distribute the product over the genital mucosa.

Summary

The ability to identify compounds and drugs that are good candidates for microbicide development is currently a significant challenge. A major part of this challenge lies in our incomplete understanding of which pathway(s) and cells in the cervical-vaginal mucosa are the main routes for sexual transmission of HIV-1. Furthermore, it is highly likely that any anti-HIV-1 micobicide product approved for vaginal use will also be used for anal intercourse. The distinct differences between the cervical-vaginal and rectal mucosae should necessitate additional preclinical studies that address the efficacy and safety of a candidate product applied to the rectal mucosa. The evidence provided by SIV and SHIV vaginal inoculation studies in NHPs are a valuable resource, but these studies do not represent the genital mucosal environment during coitus and exposure to virus in semen. Even without this understanding, the information derived from NHP and *in vitro* studies indicate that one or more of the entry inhibitor microbicides should be effective for blocking sexual transmission of R5 HIV-1 to women (Fig. 2). However, Phase III clinical testing of CS was recently halted due to preliminary results that its use may actually increase risk of HIV-1 infection [106]. The post-study evaluations of this disappointing development are still underway, so it has not yet been determined if this was an actual failure of the product or if other issues including non-compliance for product use were factors. This supports a currently held supposition that some of the first generation polyanion microbicide products will be better suited as formulations for developing combination microbicides. At least one of these third generation combination microbicides is under development [39] and more are anticipated. If stable formulations can be achieved, such a combination as a chemokine analog or gp120/gp41 ligand in a polyanion gel might prove to be highly effective. The success in developing combination microbicides will rely to a great extent on the eagerness of the different licensees to work together altruistically since a small profit margin is expected for delivering these much needed products to the developing world.

The findings and conclusions in this chapter are those of the authors and do not necessarily represent the views of the Center for Disease Control and Prevention.

References

1 Centers for Disease Control and Prevention (2005) HIV/AIDS Surveillance Report, 2004. US Department of Health and Human Services, Centers for Disease Control and Prevention, Atlanta, 1–46

2 World Health Organization (2003) The World Health Report: 2003: Shaping the Future. Chapter 3: HIV/AIDS: Confronting a Killer. World Health Organization, Geneva

3 UNAIDS (2006) 2006 Report on the Global AIDS Epidemic: Executive Summary. A UNAIDS 10th Anniversary Special Edition.: Geneva

4 Hearst N, Chen S (2004) Condom promotion for AIDS prevention in the developing world: is it working? *Stud Fam Plann* 35: 39–47

5 Shelton JD, Halperin DT, Nantulya V, Potts M, Gayle HD, Holmes KK (2004) Partner reduction is crucial for balanced "ABC" approach to HIV prevention. *BMJ* 328: 891–893

6 Stoneburner RL, Low-Beer D (2004) Population-level HIV declines and behavioral risk avoidance in Uganda. *Science* 304: 714–718

7 The Gender and Development Group (2004) Integrating Gender Issues into HIV/AIDS Programs: An Operational Guide. World Bank: Washington, D.C.

8 Cookson C (1993) WHO to concentrate HIV strategy on vaginal microbicide. *BMJ* 307: 1375–1376

9 Lange JM, Karam M, Piot P (1993) Boost for vaginal microbicides against HIV. *Lancet* 342: 1356

10 [no authors listed] (1998) Trial shows Nonoxynol 9 efficacy is questionable. *AIDS Alert* 13: 117–120

11 D'Cruz OJ, Uckun FM (2004) Clinical development of microbicides for the prevention of HIV infection. *Curr Pharm Des* 10: 315–336

12 Alliance for Microbicide Development. *Clinical Trials of Microbicides*. 2006 [accessed 16 January 2007]; available from: http://www.microbicide.org/

13 Munoz de Benito RM, Arribas Lopez JR (2006) Tenofovir disoproxil fumarate-emtricitabine coformulation for once-daily dual NRTI backbone. *Expert Rev Anti Infect Ther* 4: 523–535

14 Van Herrewege Y, Michiels J, Van Roey J, Fransen K, Kestens L, Balzarini J, Lewi P, Vanham G, Janssen P (2004) *In vitro* evaluation of nonnucleoside reverse transcriptase inhibitors UC-781 and TMC120-R147681 as human immunodeficiency virus microbicides. *Antimicrob Agents Chemother* 48: 337–339

15 Idemyor V (2005) Human immunodeficiency virus (HIV) entry inhibitors (CCR5 specific blockers) in development: are they the next novel therapies? *HIV Clin Trials* 6: 272–277

16 Strizki JM, Tremblay C, Xu S, Wojcik L, Wagner N, Gonsiorek W, Hipkin RW, Chou CC, Pugliese-Sivo C, Xiao Y et al. (2005) Discovery and characterization of vicriviroc (SCH 417690), a CCR5 antagonist with potent activity against human immunodeficiency virus type 1. *Antimicrob Agents Chemother* 49: 4911–4919

17 Trkola A, Ketas TJ, Nagashima KA, Zhao L, Cilliers T, Morris L, Moore JP, Maddon PJ, Olson WC (2001) Potent, broad-spectrum inhibition of human immunodeficiency virus type 1 by the CCR5 monoclonal antibody PRO 140. *J Virol* 75: 579–588

18 Kuritzkes DR, Jacobson J, Powderly WG, Godofsky E, DeJesus E, Haas F, Reimann KA, Larson JL, Yarbough PO, Curt V et al. (2004) Antiretroviral activity of the anti-CD4 monoclonal antibody TNX-355 in patients infected with HIV type 1. *J Infect Dis* 189: 286–291

19 Matthews T, Salgo M, Greenberg M, Chung J, DeMasi R, Bolognesi D (2004) Enfuvirtide: the first therapy to inhibit the entry of HIV-1 into host CD4 lymphocytes. *Nat Rev Drug Discov* 3: 215–225

20 Mayer KH, Peipert J, Fleming T, Fullem A, Moench T, Cu-Uvin S, Bentley M, Chesney M, Rosenberg Z (2001) Safety and tolerability of BufferGel, a novel vaginal microbicide, in women in the United States. *Clin Infect Dis* 32: 476–482

21 Chang TL, Chang CH, Simpson DA, Xu Q, Martin PK, Lagenaur LA, Schoolnik GK, Ho DD, Hillier SL, Holodniy M et al. (2003) Inhibition of HIV infectivity by a natural human isolate of *Lactobacillus jensenii* engineered to express functional two-domain CD4. *Proc Natl Acad Sci USA* 100: 11672–11677

22 Liu X, Lagenaur LA, Simpson DA, Essenmacher KP, Frazier-Parker CL, Liu Y, Tsai D, Rao SS, Hamer DH, Parks TP et al. (2006) Engineered vaginal lactobacillus strain for mucosal delivery of the human immunodeficiency virus inhibitor cyanovirin-N. *Antimicrob Agents Chemother* 50: 3250–3259

23 Pusch O, Kalyanaraman R, Tucker LD, Wells JM, Ramratnam B, Boden D (2006) An anti-HIV microbicide engineered in commensal bacteria: secretion of HIV-1 fusion inhibitors by lactobacilli. *AIDS* 20: 1917–1922
24 Bax R, K. Douville, D. McCormick, M. Rosenberg, J. Higgins, M. Dowden (2002) Microbicides-evaluating multiple formulations of C31G. *Contraception* 66: 365–368
25 Rusert P, Kuster H, Joos B, Misselwitz B, Gujer C, Leemann C, Fischer M, Stiegler G, Katinger H, Olson WC et al. (2005) Virus isolates during acute and chronic human immunodeficiency virus type 1 infection show distinct patterns of sensitivity to entry inhibitors. *J Virol* 79: 8454–8469
26 Boyd MR, Gustafson KR, McMahon JB, Shoemaker RH, O'Keefe BR, Mori T, Gulakowski RJ, Wu L, Rivera MI, Laurencot CM et al. (1997) Discovery of cyanovirin-N, a novel human immunodeficiency virus-inactivating protein that binds viral surface envelope glycoprotein gp120: potential applications to microbicide development. *Antimicrob Agents Chemother* 41: 1521–1530
27 Joglekar NS, Joshi SN, Navlakha SN, Katti UR, Mehendale SM (2006) Acceptability of Praneem polyherbal vaginal tablet among HIV uninfected women and their male partners in Pune, India – Phase I study. *Indian J Med Res* 123: 547–552
28 Edwards JN, Morris HB (1985) Langerhans' cells and lymphocyte subsets in the female genital tract. *Br J Obstet Gynaecol* 92: 974–982
29 Miller CJ, Shattock RJ (2003) Target cells in vaginal HIV transmission. *Microbes Infect* 5: 59–67
30 Hu J, Gardner MB, Miller CJ (2000) Simian immunodeficiency virus rapidly penetrates the cervicovaginal mucosa after intravaginal inoculation and infects intraepithelial dendritic cells. *J Virol* 74: 6087–6095
31 Cummins JE, Christensen L, Lennox JL, Bush TJ, Wu Z, Malamud D, Evans-Strickfaden T, Siddig A, Caliendo AM, Hart CE et al. (2006) Mucosal innate immune factors in the female genital tract are associated with vaginal HIV-1 shedding independent of plasma viral load. *AIDS Res Hum Retroviruses* 22: 788–795
32 Hart CE, Lennox JL, Pratt-Palmore M, Wright TC, Schinazi RF, Evans-Strickfaden T, Bush TJ, Schnell C, Conley LJ, Clancy KA et al. (1999) Correlation of human immunodeficiency virus type 1 RNA levels in blood and the female genital tract. *J Infect Dis* 179: 871–882
33 Moulard M, Lortat-Jacob H, Mondor I, Roca G, Wyatt R, Sodroski J, Zhao L, Olson W, Kwong PD, Sattentau QJ (2000) Selective interactions of polyanions with basic surfaces on human immunodeficiency virus type 1 gp120. *J Virol* 74: 1948–1960
34 Margolis M, Shattock R (2006) Selective transmission of CCR5-utilizing HIV-1: the 'gatekeeper' problem resolved? *Nat Rev Microbiol* 4: 312–317
35 Scordi-Bello IA, Mosoian A, He C, Chen Y, Cheng Y, Jarvis GA, Keller MJ, Hogarty K, Waller DP, Profy AT et al. (2005) Candidate sulfonated and sulfated topical microbicides: comparison of anti-human immunodeficiency virus activities and mechanisms of action. *Antimicrob Agents Chemother* 49: 3607–3615
36 Anderson RA, Feathergill KA, Diao XH, Cooper MD, Kirkpatrick R, Herold BC, Doncel GF, Chany CJ, Waller DP, Rencher WF et al. (2002) Preclinical evaluation of sodium cellulose sulfate (Ushercell) as a contraceptive antimicrobial agent. *J Androl* 23: 426–438
37 Neurath AR, Strick N, Li YY (2002) Anti-HIV-1 activity of anionic polymers: a comparative study of candidate microbicides. *BMC Infect Dis* 2: 27–38
38 Dezzutti CS, James VN, Ramos A, Sullivan ST, Siddig A, Bush TJ, Grohskopf LA, Paxton L, Subbarao S, Hart CE (2004) *In vitro* comparison of topical microbicides for prevention of human immunodeficiency virus type 1 transmission. *Antimicrob Agents Chemother* 48: 3834–3844
39 Fernandez-Romero JA, Thorn M, Turville SG, Titchen K, Sudol K, Li J, Miller T, Robbiani M, Maguire RA, Buckheit RW, Jr et al. (2007) Carrageenan/MIV-150 (PC-815), a combination microbicide. *Sex Transm Dis* 34: 9–14
40 Fletcher PS, Wallace GS, Mesquita PM, Shattock RJ (2006) Candidate polyanion microbicides inhibit HIV-1 infection and dissemination pathways in human cervical explants. *Retrovirology* 3: 46–57
41 Jiang S, Lin K, Zhang L, Debnath AK (1999) A screening assay for antiviral compounds targeted to the HIV-1 gp41 core structure using a conformation-specific monoclonal antibody. *J Virol Methods* 80: 85–96
42 Neurath AR, Strick N, Jiang S, Li YY, Debnath AK (2002) Anti-HIV-1 activity of cellulose acetate phthalate: synergy with soluble CD4 and induction of "dead-end" gp41 six-helix bundles. *BMC*

Infect Dis 2: 6–18

43 Greenhead P, Hayes P, Watts PS, Laing KG, Griffin GE, Shattock RJ (2000) Parameters of human immunodeficiency virus infection of human cervical tissue and inhibition by vaginal virucides. *J Virol* 74: 5577–5586

44 Weber J, Nunn A, O'Connor T, Jeffries D, Kitchen V, McCormack S, Stott J, Almond N, Stone A, Darbyshire J (2001) 'Chemical condoms' for the prevention of HIV infection: evaluation of novel agents against SHIV(89.6PD) *in vitro* and *in vivo*. *AIDS* 15: 1563–1568

45 Keller MJ, Zerhouni-Layachi B, Cheshenko N, John M, Hogarty K, Kasowitz A, Goldberg CL, Wallenstein S, Profy AT, Klotman ME et al. (2006) PRO 2000 gel inhibits HIV and herpes simplex virus infection following vaginal application: a double-blind placebo-controlled trial. *J Infect Dis* 193: 27–35

46 King JB, Stickler DJ (1991) An assessment of antiseptic bladder washout solutions using a physical model of the catheterized bladder. *J Hosp Infect* 18: 179–190

47 Pearce-Pratt R, Phillips DM (1996) Sulfated polysaccharides inhibit lymphocyte-to-epithelial transmission of human immunodeficiency virus-1. *Biol Reprod* 54: 173–182

48 Lu H, Zhao Q, Wallace G, Liu S, He Y, Shattock R, Neurath AR, Jiang BS (2006) Cellulose acetate 1,2-benzenedicarboxylate inhibits infection by cell-free and cell-associated primary HIV-1 isolates. *AIDS Res Hum Retroviruses* 22: 411–418

49 Otten RA, Adams DR, Kim CN, Jackson E, Pullium JK, Lee K, Grohskopf LA, Monsour M, Butera S, Folks TM (2005) Multiple vaginal exposures to low doses of R5 simian-human immunodeficiency virus: strategy to study HIV preclinical interventions in nonhuman primates. *J Infect Dis* 191: 164–173

50 Jiang YH, Emau P, Cairns JS, Flanary L, Morton WR, McCarthy TD, Tsai CC (2005) SPL7013 gel as a topical microbicide for prevention of vaginal transmission of SHIV89.6P in macaques. *AIDS Res Hum Retroviruses* 21: 207–213

51 Herold BC, Scordi-Bello I, Cheshenko N, Marcellino D, Dzuzelewski M, Francois F, Morin R, Casullo VM, Anderson RA, Chany C 2nd et al. (2002) Mandelic acid condensation polymer: novel candidate microbicide for prevention of human immunodeficiency virus and herpes simplex virus entry. *J Virol* 76: 11236–11244

52 Coetzer M, Cilliers T, Ping LH, Swanstrom R, Morris L (2006) Genetic characteristics of the V3 region associated with CXCR4 usage in HIV-1 subtype C isolates. *Virology* 356: 95–105

53 Fouchier RA, Groenink M. Koostra NA, Tersmette M, Huisman HG, Miedema F, Schuitemaker H (1992) Phenotype-associated sequence variation in the third variable domain of the human immunodeficiency virus type 1 gp120 molecule. *J Virol* 66: 3183–3187

54 Doms RW, Trono D (2000) The plasma membrane as a combat zone in the HIV battlefield. *Genes Dev* 14: 2677–2688

55 Vives RR, Imberty A, Sattentau QJ, Lortat-Jacob H (2005) Heparan sulfate targets the HIV-1 envelope glycoprotein gp120 coreceptor binding site. *J Biol Chem* 280: 21353–21357

56 Shattock RJ, Doms RW (2002) AIDS models: microbicides could learn from vaccines. *Nat Med* 8: 425

57 Roben P, Moore JP, Thali M, Sodroski J, Barbas CF 3rd, Burton DR (1994) Recognition properties of a panel of human recombinant Fab fragments to the CD4 binding site of gp120 that show differing abilities to neutralize human immunodeficiency virus type 1. *J Virol* 68: 4821–4828

58 Saphire EO, Parren PW, Pantophlet R, Zwick MB, Morris GM, Rudd PM, Dwek RA, Stanfield RL, Burton DR, Wilson IA (2001) Crystal structure of a neutralizing human IGG against HIV-1: a template for vaccine design. *Science* 293: 1155–1159

59 Sanders RW, Venturi M, Schiffner L, Kalyanaraman R, Katinger H, Lloyd KO, Kwong PD, Moore JP (2002) The mannose-dependent epitope for neutralizing antibody 2G12 on human immunodeficiency virus type 1 glycoprotein gp120. *J Virol* 76: 7293–7305

60 Scanlan CN, Pantophlet R, Wormald MR, Ollmann Saphire E, Stanfield R, Wilson IA, Katinger H, Dwek RA, Rudd PM, Burton DR (2002) The broadly neutralizing anti-human immunodeficiency virus type 1 antibody 2G12 recognizes a cluster of alpha12 mannose residues on the outer face of gp120. *J Virol* 76: 7306–7321

61 Parren PW, Burton DR (2001) The antiviral activity of antibodies *in vitro* and *in vivo*. *Adv Immunol* 77: 195–262

62 Zwick MB, Wang M, Poignard P, Stiegler G, Katinger H, Burton DR, Parren PW (2001) Neutralization synergy of human immunodeficiency virus type 1 primary isolates by cocktails of broadly neutralizing antibodies. *J Virol* 75: 12198–12208

63 Veazey RS, Shattock RJ, Pope M, Kirijan JC, Jones J, Hu Q, Ketas T, Marx PA, Klasse PJ, Burton DR et al. (2003) Prevention of virus transmission to macaque monkeys by a vaginally applied monoclonal antibody to HIV-1 gp120. *Nat Med* 9: 343–346
64 Mascola JR, Stiegler G, VanCott TC, Katinger H, Carpenter CB, Hanson CE, Beary H, Hayes D, Frankel SS, Birx DL et al. (2000) Protection of macaques against vaginal transmission of a pathogenic HIV-1/SIV chimeric virus by passive infusion of neutralizing antibodies. *Nat Med* 6: 207–210
65 Frank I, Pope M (2002) The enigma of dendritic cell-immunodeficiency virus interplay. *Curr Mol Med* 2: 229–248
66 Huskens D, Van Laethem K, Vermeire K, Balzarini J, Schols D (2007) Resistance of HIV-1 to the broadly HIV-1-neutralizing, anti-carbohydrate antibody 2G12. *Virology* 360: 294–304
67 Shattock RJ, Griffin GE, Gorodeski GI (2000) *In vitro* models of mucosal HIV transmission. *Nat Med* 6: 607–608
68 Li M, Salazar-Gonzalez JF, Derdeyn CA, Morris L, Williamson C, Robinson JE, Decker JM, Li Y, Salazar MG, Polonis VR et al. (2006) Genetic and neutralization properties of subtype C human immunodeficiency virus type 1 molecular env clones from acute and early heterosexually acquired infections in Southern Africa. *J Virol* 80: 11776–11790
69 Zhang MY, Xiao X, Sidorov IA, Choudhry V, Cham F, Zhang PF, Bouma P, Zwick M, Choudhary A, Montefiori DC et al. (2004) Identification and characterization of a new cross-reactive human immunodeficiency virus type 1-neutralizing human monoclonal antibody. *J Virol* 78: 9233–9242
70 Dove A (2002) Uncorking the biomanufacturing bottleneck. *Nat Biotechnol* 20: 777–779
71 Lederman MM, Offord RE, Hartley O (2006) Microbicides and other topical strategies to prevent vaginal transmission of HIV. *Nat Rev Immunol* 6: 371–382
72 Gallaher WR, Ball JM, Garry RF, Martin-Amedee AM, Montelaro RC (1995) A general model for the surface glycoproteins of HIV and other retroviruses. *AIDS Res Hum Retroviruses* 11: 191–202
73 Leonard CK, Spellman MW, Riddle L, Harris RJ, Thomas JN, Gregory TJ (1990) Assignment of intrachain disulfide bonds and characterization of potential glycosylation sites of the type 1 recombinant human immunodeficiency virus envelope glycoprotein (gp120) expressed in Chinese hamster ovary cells. *J Biol Chem* 265: 10373–10382
74 Balzarini J (2006) Inhibition of HIV entry by carbohydrate-binding proteins. *Antiviral Res* 71: 237–247
75 Balzarini J, Van Laethem K, Hatse S, Froeyen M, Peumans W, Van Damme E, Schols D (2005) Carbohydrate-binding agents cause deletions of highly conserved glycosylation sites in HIV GP120: a new therapeutic concept to hit the achilles heel of HIV. *J Biol Chem* 280: 41005–41014
76 Dey B, Lerner DL, Lusso P, Boyd MR, Elder JH, Berger EA (2000) Multiple antiviral activities of cyanovirin-N: blocking of human immunodeficiency virus type 1 gp120 interaction with CD4 and coreceptor and inhibition of diverse enveloped viruses. *J Virol* 74: 4562–4569
77 Tsai CC, Emau P, Jiang Y, Agy MB, Shattock RJ, Schmidt A, Morton WR, Gustafson KR, Boyd MR (2004) Cyanovirin-N inhibits AIDS virus infections in vaginal transmission models. *AIDS Res Hum Retroviruses* 20: 11–18
78 Balzarini J, Van Laethem K, Peumans WJ, Van Damme EJ, Bolmstedt A, Gago F, Schols D (2006) Mutational pathways, resistance profile, and side effects of cyanovirin relative to human immunodeficiency virus type 1 strains with N-glycan deletions in their gp120 envelopes. *J Virol* 80: 8411–8421
79 Kwong PD, Doyle ML, Casper DJ, Cicala C, Leavitt SA, Majeed S, Steenbeke TD, Venturi M, Chaiken I, Fung M et al. (2002) HIV-1 evades antibody-mediated neutralization through conformational masking of receptor-binding sites. *Nature* 420: 678–682
80 Geijtenbeek TB, Kwon DS, Torensma R, van Vliet SJ, van Duijnhoven GC, Middel J, Cornelissen IL, Nottet HS, KewalRamani VN, Littman DR et al. (2000) DC-SIGN, a dendritic cell-specific HIV-1-binding protein that enhances trans-infection of T cells. *Cell* 100: 587–597
81 Curtis BM, Scharnowske S, Watson AJ (1992) Sequence and expression of a membrane-associated C-type lectin that exhibits CD4-independent binding of human immunodeficiency virus envelope glycoprotein gp120. *Proc Natl Acad Sci USA* 89: 8356–8360
82 Hong PW, Flummerfelt KB, de Parseval A, Gurney K, Elder JH, Lee B (2002) Human immunodeficiency virus envelope (gp120) binding to DC-SIGN and primary dendritic cells is carbohydrate dependent but does not involve 2G12 or cyanovirin binding sites: implications for structural analyses of gp120-DC-SIGN binding. *J Virol* 76: 12855–12865

83 Balzarini J, Van Herrewege Y, Vermeire K, Vanham G, Schols D (2007) Carbohydrate-binding agents efficiently prevent dendritic cell-specific intercellular adhesion molecule-3-grabbing non-integrin (DC-SIGN)-directed HIV-1 transmission to T lymphocytes. *Mol Pharmacol* 71: 3–11
84 Snyder GA, Ford J, Torabi-Parizi P, Arthos JA, Schuck P, Colonna M, Sun PD (2005) Characterization of DC-SIGN/R interaction with human immunodeficiency virus type 1 gp120 and ICAM molecules favors the receptor's role as an antigen-capturing rather than an adhesion receptor. *J Virol* 79: 4589–4598
85 Liu H, Hladik F, Andrus T, Sakchalathorn P, Lentz GM, Fialkow MF, Corey L, McElrath MJ, Zhu T (2005) Most DC-SIGNR transcripts at mucosal HIV transmission sites are alternatively spliced isoforms. *Eur J Hum Genet* 13: 707–715
86 Liu H, Carrington M, Wang C, Holte S, Lee J, Greene B, Hladik F, Koelle DM, Wald A, Kurosawa K et al. (2006) Repeat-region polymorphisms in the gene for the dendritic cell-specific intercellular adhesion molecule-3-grabbing nonintegrin-related molecule: effects on HIV-1 susceptibility. *J Infect Dis* 193: 698–702
87 Jacobson JM, Israel RJ, Lowy I, Ostrow NA, Vassilatos LS, Barish M, Tran DN, Sullivan BM, Ketas TJ, O'Neill TJ et al. (2004) Treatment of advanced human immunodeficiency virus type 1 disease with the viral entry inhibitor PRO 542. *Antimicrob Agents Chemother* 48: 423–429
88 Wang HG, Williams RE, Lin PF (2004) A novel class of HIV-1 inhibitors that targets the viral envelope and inhibits CD4 receptor binding. *Curr Pharm Des* 10: 1785–1793
89 Si Z, Madani N, Cox JM, Chruma JJ, Klein JC, Schon A, Phan N, Wang L, Biorn AC, Cocklin S et al. (2004) Small-molecule inhibitors of HIV-1 entry block receptor-induced conformational changes in the viral envelope glycoproteins. *Proc Natl Acad Sci USA* 101: 5036–5041
90 Ji H, Bracken C, Lu M (2000) Buried polar interactions and conformational stability in the simian immunodeficiency virus (SIV) gp41 core. *Biochemistry* 39: 676–685
91 Guo Q, Ho HT, Dicker I, Fan L, Zhou N, Friborg J, Wang T, McAuliffe BV, Wang HG, Rose RE et al. (2003) Biochemical and genetic characterizations of a novel human immunodeficiency virus type 1 inhibitor that blocks gp120-CD4 interactions. *J Virol* 77: 10528–10536
92 Lin PF, Blair W, Wang T, Spicer T, Guo Q, Zhou N, Gong YF, Wang HG, Rose R, Yamanaka G et al. (2003) A small molecule HIV-1 inhibitor that targets the HIV-1 envelope and inhibits CD4 receptor binding. *Proc Natl Acad Sci USA* 100: 11013–11018
93 Hu Q, Frank I, Williams V, Santos JJ, Watts P, Griffin GE, Moore JP, Pope M, Shattock RJ (2004) Blockade of attachment and fusion receptors inhibits HIV-1 infection of human cervical tissue. *J Exp Med* 199: 1065–1075
94 Veazey RS, Klasse PJ, Schader SM, Hu Q, Ketas TJ, Lu M, Marx PA, Dufour J, Colonno RJ, Shattock RJ et al. (2005) Protection of macaques from vaginal SHIV challenge by vaginally delivered inhibitors of virus-cell fusion. *Nature* 438: 99–102
95 Alkhatib G, Combadiere C, Broder CC, Feng Y, Kennedy PE, Murphy PM, Berger EA (1996) CC CKR5: a RANTES, MIP-1alpha, MIP-1beta receptor as a fusion cofactor for macrophage-tropic HIV-1. *Science* 272: 1955–1958
96 Bleul CC, Farzan M, Choe H, Parolin C, Clark-Lewis I, Sodroski J, Springer TA (1996) The lymphocyte chemoattractant SDF-1 is a ligand for LESTR/fusin and blocks HIV-1 entry. *Nature* 382: 829–833
97 Choe H, Farzan M, Sun Y, Sullivan N, Rollins B, Ponath PD, Wu L, Mackay CR, LaRosa G, Newman W et al. (1996) The beta-chemokine receptors CCR3 and CCR5 facilitate infection by primary HIV-1 isolates. *Cell* 85: 1135–1148
98 Deng H, Liu R, Ellmeier W, Choe S, Unutmaz D, Burkhart M, Di Marzio P, Marmon S, Sutton RE, Hill CM et al. (1996) Identification of a major co-receptor for primary isolates of HIV-1. *Nature* 381: 661–666
99 Dragic T, Litwin V, Allaway GP, Martin SR, Huang Y, Nagashima KA, Cayanan C, Maddon PJ, Koup RA, Moore JP et al. (1996) HIV-1 entry into $CD4^+$ cells is mediated by the chemokine receptor CC-CKR-5. *Nature* 381: 667–673
100 Berger EA, Murphy PM, Farber JM (1999) Chemokine receptors as HIV-1 coreceptors: roles in viral entry, tropism, and disease. *Annu Rev Immunol* 17: 657–700
101 Clapham PR, McKnight A (2002) Cell surface receptors, virus entry and tropism of primate lentiviruses. *J Gen Virol* 83: 1809–1829
102 Hartley O, Gaertner H, Wilken J, Thompson D, Fish R, Ramos A, Pastore C, Dufour B, Cerini F, Melotti A et al. (2004) Medicinal chemistry applied to a synthetic protein: development of highly potent HIV entry inhibitors. *Proc Natl Acad Sci USA* 101: 16460–16465
103 Amara A, Gall SL, Schwartz O, Salamero J, Montes M, Loetscher P, Baggiolini M, Virelizier JL,

Arenzana-Seisdedos F (1997) HIV coreceptor downregulation as antiviral principle: SDF-1alpha-dependent internalization of the chemokine receptor CXCR4 contributes to inhibition of HIV replication. *J Exp Med* 186: 139–146

104 Lederman MM, Veazey RS, Offord R, Mosier DE, Dufour J, Mefford M, Piatak M, Jr, Lifson JD, Salkowitz JR, Rodriguez B et al. (2004) Prevention of vaginal SHIV transmission in rhesus macaques through inhibition of CCR5. *Science* 306: 485–487

105 Shen DM, Shu M, Willoughby CA, Shah S, Lynch CL, Hale JJ, Mills SG, Chapman KT, Malkowitz L, Springer MS et al. (2004) Antagonists of human CCR5 receptor containing 4-(pyrazolyl)piperidine side chains. Part 2: Discovery of potent, selective, and orally bioavailable compounds. *Bioorg Med Chem Lett* 14: 941–945

106 International AIDS Society (2007) *International AIDS Society Statement on Announcement that Two Phase III Trials of HIV Microbiocide Candidate Ushercell Have Benn Halted.* [accessed 2007 Mar 2]; available from http://www.iasociety.org/images/upload/1263.pdf

Entry Inhibitors in HIV Therapy
Edited by Jacqueline D. Reeves and Cynthia A. Derdeyn

Entry inhibition of HIV-1 subtype C isolates

Lynn Morris[1], Mia Coetzer[1], Elin S. Gray[1], Tonie Cilliers[1], Kabamba B. Alexandre [1], Penny L. Moore[1] and James M. Binley[2]

[1] *AIDS Virus Research Unit, National Institute for Communicable Diseases, Private Bag X4, Sandringham 2131, Johannesburg, South Africa*
[2] *Torrey Pines Institute for Molecular Studies, 3550 General Atomics Court, San Diego, CA 92121, USA*

Genetic diversity of the HIV envelope glycoprotein and coreceptor usage

The HIV shows an extraordinary degree of genetic diversity that in rare cases impacts on the efficacy of currently available anti-retroviral therapies [1]. There is less information on how genetic variability might affect the efficacy of a newer class of anti-retrovirals, the entry or fusion inhibitors. This group comprises a diverse collection of compounds that target both viral and host cell components blocking virus attachment and/or fusion and preventing infection and viral integration. Available data suggests that almost all genetic subtypes of HIV-1 engage the CD4 and coreceptor molecules. Thus entry inhibitors that target cellular proteins are likely to be equally efficacious across genetic subtypes. However, the efficacy of those that target the viral envelope glycoprotein is likely to be more impacted, given that as the envelope gene is the most variable of all HIV genes, showing up to 30% difference within HIV-1 and up to 55% between HIV-1 and HIV-2 (www.hiv.lanl.gov). Most entry inhibitors have been designed and tested based on HIV-1 subtype B viruses, and would therefore be expected to be most effective against viruses of this genetic subtype. There are few studies that have specifically explored the phenotypic sensitivity of different HIV subtypes to entry inhibitors. Since HIV-1 subtype C is now the most prevalent subtype globally, causing explosive epidemics in southern Africa, Ethiopia, India and China, we focus our review primarily on viruses of this subtype.

HIV-1 subtype C viruses predominantly use the CCR5 coreceptor for viral entry with CXCR4 usage reported less frequently compared to subtype B [2–5]. Among 231 HIV-1 subtype C viral isolates in our laboratory, only 10% were found to use CXCR4, and this was significantly correlated with a CD4 count of <200 cells/μl [6], similar to subtype B [7, 8]. Thus, over 80% of patients with clinically defined subtype C-induced AIDS still harbor viruses that use CCR5 [4]. The reasons for this are unclear, but could be related to host factors, such as elevated expression levels of the CCR5 coreceptor favoring expansion of R5 viruses [9], or a shorter time to AIDS in countries with poor health care infrastructure limiting the opportunity for CXCR4-using viruses to develop. There may also be virological factors restricting envelope evolution. Understanding HIV-1 coreceptor usage is now particularly relevant in the era of entry inhibitors. Whether or not subtype C will be more amenable to CCR5 antagonists, for instance, will depend on whether coreceptor usage changes as the epidemic matures.

Analysis of the V3 region, which is a major determinant of coreceptor usage, indicates that CXCR4 usage among subtype C viruses is associated with genetic changes, including increased numbers of positively charged amino acids, length variations and loss of a glycosylation site, similar to that reported for subtype B [6, 10]. However, in contrast to subtype B, where both CCR5- and CXCR4-using viruses contain a consensus GPGR crown motif, subtype C CXCR4 viruses frequently have positively charged substitutions in the GPGQ crown [6]. Whether the lack of a positively charged amino acid (arginine, R) in the V3 crown limits the ability of subtype C viruses to undergo a coreceptor switch requires further study, but may explain the low frequency of CXCR4-using viruses in subtype C.

Inhibiting entry of HIV-1 subtype C viruses

Over the last two decades a large number of entry inhibitors have been developed, targeting the three stages of viral entry, i.e., CD4 binding, interaction with CCR5 and/or CXCR4 and fusion. Available *in vitro* data suggest that almost all are highly effective against HIV-1 subtype C viruses [4, 11, 12]. We found that PRO542 (also known as CD4-IgG2), a recombinant antibody fused with four copies of CD4, as well as soluble CD4 (sCD4), both of which prevent CD4-gp120 interactions, inhibit HIV-1 subtype C infections independent of coreceptor use (Fig. 1). In contrast, CCR5 coreceptor antagonists, including RANTES (a chemokine) and PRO140 [an anti-CCR5 monoclonal antibody (mAb)] fully inhibited R5 isolates, but were largely ineffective against isolates that use CXCR4. Variability in the levels of inhibition of R5X4 viruses may be due to the presence of mixed populations of R5 and X4 viruses. The small-molecule CXCR4 inhibitor AMD3100 is able to inhibit all R5X4 and X4 isolates, with no activity against the R5 isolates. Enfuvirtide (ENF, also know as T-20), a 36-amino acid peptide that prevents virus fusion by binding to the HR-1 region in gp41, was effective against subtype C viruses irrespective of coreceptor usage. However, other studies have shown that higher concentrations of ENF are needed to inhibit subtype B R5 isolates perhaps due to more rapid fusion rates [13, 14].

A potential obstacle to using coreceptor inhibitors is the emergence of viral variants that use coreceptors other than CCR5 or CXCR4 for cell entry. Although the use of alternate coreceptors appears to be a rare event with scant evidence to suggest that they play major a role *in vivo* [15], such isolates could pose a threat to the success of coreceptor inhibitors in the clinic. In subtype C,

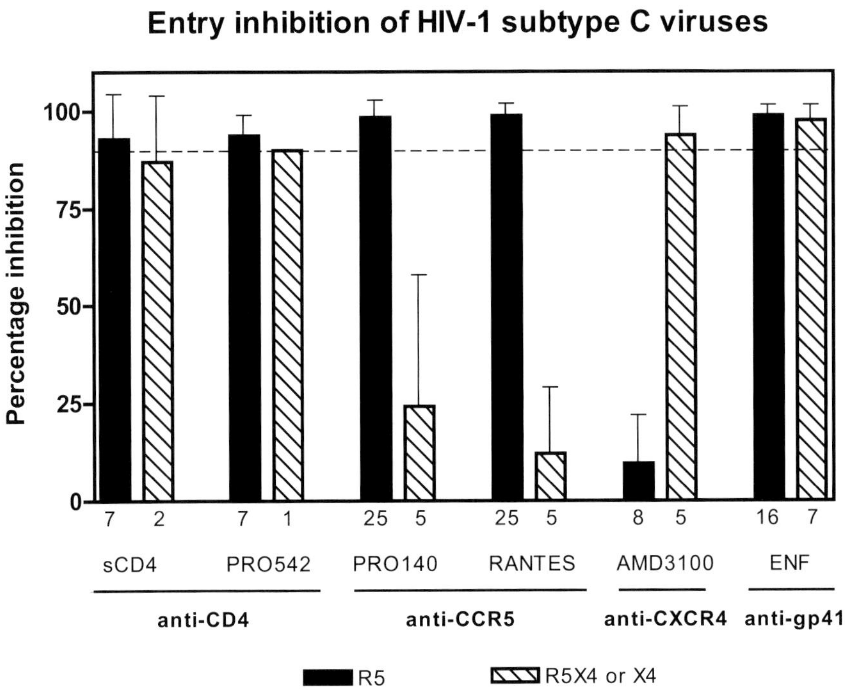

Figure 1. Sensitivity of HIV-1 subtype C isolates to all three classes of entry inhibitors. Results are shown as the mean percentage inhibition plus standard deviation for isolates able to use CCR5 (black bars) or those able to use CXCR4 (R5X4 and X4) (patterned bars). The number of isolates tested is shown at the bottom of each bar. sCD4 and PRO540 were both tested at 50 µg/ml, PRO140 was tested at 167 nM, RANTES at 64 nM, AMD3100 at 500 nM and ENF at 1 µg/ml. The dotted line shows 90% inhibition. Data are from [4, 11, 12], plus unpublished data. Figure adapted from the PhD thesis of Tonie Cilliers "Effectiveness of entry inhibitors on HIV-1 subtype C viruses", University of the Witwatersrand, Johannesburg, South Africa. 2005.

viruses able to use CXCR6, BOB/GPR15, CCR3 and CCR8 on transfected cell lines have been described [3, 4, 16]. In one case, the use of CCR8 was confirmed in primary cells and in another the receptor could not be identified [16]. Given that few HIV-1 isolates are tested for alternate coreceptor use, such receptors may be more commonly used than current data indicate [17, 18].

Natural and induced resistance to entry inhibitors

As with other anti-retrovirals, resistance to entry inhibitors is an inevitable consequence of their use, particularly if they are used as monotherapy. However, unlike the emergence of resistance to reverse transcriptase and protease inhibitors, which tends to be associated with signature mutations, resistance in the envelope gene is considerably more complex [19]. Escape from the

CCR5 inhibitors SCH-D and SCH-C, for example, was associated with multiple changes in gp120 that differed between patients [20]. While one might expect that resistance to CCR5 inhibitors may select for X4 variants, it appears that the use of CCR5 in an inhibitor-insensitive way is a more common pathway to resistance against this class of drugs [21, 22]. In one case where coreceptor switching was seen, the authors speculated that this was due to preexisting X4 variants and not the result of treatment with the CCR5 inhibitor, maraviroc [23].

Natural resistance to the CD4 inhibitor BMS-378806 among non-B subtypes was predicted based on the genetic analysis of sequences from subtypes A–G [24]. This molecule binds to viral gp120 and prevents interaction with cellular CD4 receptors, and resistance has been mapped to at least nine loci around the CD4-binding pocket [25]. Background genotypic resistance to BMS-378806 among non-B HIV-1 viruses was found to be higher than in subtype B [24], an observation that is supported by limited phenotypic data [25]. Whether or not this molecule and its derivative BMS-488043 [26] will show reduced efficacy against non-B subtypes awaits clinical testing.

Subtype C viruses show significant genetic variation from subtype B in the HR-2 region that was used to design the fusion inhibitor ENF. Of the 36 amino acids that constitute ENF, 10–16 residues differ among subtype C sequences, without any apparent compromise in its ability to inhibit subtype C viruses, at least *in vitro* [27]. Resistance to ENF is largely determined in the HR-1 region around the GIV motif. This is highly conserved among all HIV-1 genetic subtypes and all are predicted to be fully sensitive to ENF, for which there is some experimental evidence [28]. Only HIV-2 shows reduced sensitivity to ENF, probably because of the large degree of genetic variation in the HR-2 region [29]. Induced resistance to ENF by *in vitro* culture of subtype C isolates in the presence of increasing concentrations of the drug indicates almost identical signature mutations to HIV-1 subtype B, suggesting that the mechanism of action of ENF is the same for subtype C viruses [27].

Predicting HIV-1 subtype C coreceptor usage

Given that coreceptor inhibitors are showing promise in human clinical trials, there is a need to monitor coreceptor use within a patient to maximize the efficacy of therapy. A number of predictive algorithms have been developed based on characteristics within the V3 region associated with CCR5 and CXCR4 usage. The most basic and still one of the most reliable algorithms for predicting coreceptor usage is the “11/25” or “charge rule”, which classifies isolates as X4 if they have a positively charged amino acid (R, K or H) at positions 11 and/or 25 [30]. Other approaches, including the use of neural networks, a support vector machine (SVM), the Briggs method and a position-specific scoring matrix (PSSM), often show improved sensitivity and specificity, but are more complex to use [31]. These prediction methods also have some limitations in that they do not take regions outside of V3 into account, and that minor populations might be missed. In addition, these algorithms cannot distinguish true dual-tropic from mixtures of CCR5 and CXCR4 using variants. Nevertheless, the ability to use sequence data to predict coreceptor usage offers significant advantages over phenotyping, which is costly and labor intensive. However, all of these algorithms were developed using HIV-1 subtype B datasets, and one of the limitations in developing algorithms for other subtypes is the paucity of sequence data from isolates of non-B subtypes of known coreceptor specificity. A C-PSSM that was developed using subtype C sequence datasets showed increased sensitivity for detecting CXCR4-using variants over a B-PSSM, although the specificity for detecting CCR5 variants was slightly compromised [32]. Refinements of these algorithms, together with knowledge of mutations associated with resistance to each of the entry inhibitors, will facilitate improved clinical application of these agents.

Inhibiting entry by mAbs

mAbs inhibit HIV entry by blocking virus binding to CD4 and/or coreceptor or post-receptor binding refolding events. Although neutralizing antibodies (nAbs) against envelope are frequently less potent than inhibitors that target cellular receptors, the lack of safety concerns in targeting the virus make them more attractive for clinical use. Cocktails of these mAbs have been shown to be capable of effectively blocking infection via intravenous or mucosal routes in animal studies [33–37]. However, unlike the inhibitors of structural and regulatory virus components, nAbs have little potency against established or chronic infection [38]. nAbs are, therefore, likely to have their greatest impact in the early stages of infection, e.g., when formulated as a preventative mucosal microbicide or in post-exposure prophylaxis.

Advances in single-round pseudovirus assays for measuring neutralization have enabled expanded investigation of nAb cross-reactivity. One study investigated the reactivity of several known neutralizing mAbs against 90 viruses from different genetic subtypes [39]. Anti-gp120 mAb IgG1b12, directed against an epitope overlapping the CD4-binding site, neutralized 50% of viruses, including some from almost every subtype, albeit with a lower percentage against non-B (39%) than B viruses (72%). 2G12, directed against a high mannose epitope of gp120 neutralized 41% of viruses, but none from subtypes C or E. mAb 4E10, directed to the C terminus of the gp41 ectodomain, neutralized all viruses with moderate potency. mAb 2F5, directed against an epitope adjacent to that of 4E10, neutralized 67% of isolates, but none from subtype C (Fig. 2).

The observation that only two of the four main neutralizing mAbs effectively neutralized subtype C viruses is perhaps a reflection, in part, of the fact that the mAbs were all isolated from tissues donated by subtype B-infected donors. Sequence differences explain some of the lack of cross-reactivity with

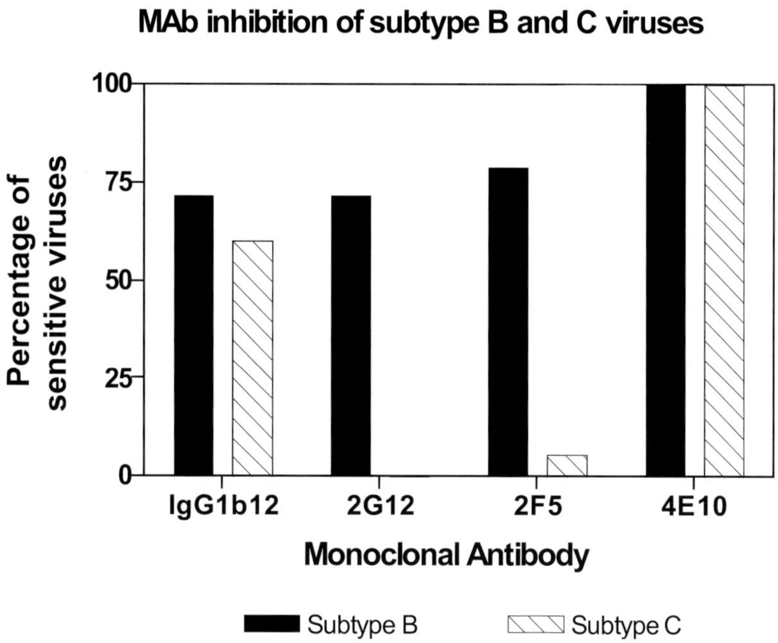

Figure 2. Insensitivity of HIV-1 subtype C viruses to 2G12 and 2F5 neutralizing mAbs. The percentage of subtype B and subtype C viruses sensitive to neutralization by the mAbs IgG1b12, 2G12, 2F5 and 4E10 is shown. mAbs were tested at 50 μg/ml. Data taken from [39, 40, 65].

subtype C viruses [39, 40]. The minimal requirement for 2F5 neutralization is the motif "DKW" at positions 664–666 in gp41, with all viruses having a substitution at K665 showing resistance to 2F5. In the Los Alamos database, the K665 is preserved in only 12% of subtype C viruses, and is disproportionally enriched or absent in certain regional subtype C viruses, suggesting a geographic lineage association with 2F5 susceptibility [39]. Regarding 2G12, the lack of efficacy against subtype C viruses is consistent with the lack of a glycan at position 295 at the N-terminal base of the V3 loop (absent in >83% of subtype C viruses). However, reintroduction of this glycan by site-directed mutagenesis restored neutralization sensitivity in only one of three subtype C isolates analyzed [41], suggesting that there may be structural differences between subtype B and C envelopes that precludes the formation of the mannose cluster that constitutes the 2G12 epitope.

Features of transmitted variants

An understanding of which variants from among the myriad of circulating HIV-1 strains in an infected individual are most commonly transmitted, and their susceptibility to various entry inhibitors, is clearly of paramount importance in research aimed at using entry inhibitors for prophylaxis. Studies in an African subtype C cohort of discordant heterosexual couples suggested that transmitted viruses are particularly neutralization-sensitive, due to shortened variable loops and reduced glycosylation [42]. However, a study of a homosexual subtype B cohort was unable to show that transmitted viruses were more sensitive than the donor virus [43]. In another study, comparing subtypes A and B, shorter V1/V2 loops and fewer N-linked glycosylation sites was a hallmark of subtype A heterosexual transmission, but not subtype B transmission, regardless of whether this occurred through hetero- or homosexual contact [44]. This suggests that the distinction between selection of subtypes A and C *versus* subtype B viruses is lineage specific, not transmission-mode dependent. A recent study of mother to infant transmission in Nairobi (including subjects infected with subtypes A and C, as well as D/A and C/D recombinants) indicated that vertically transmitted viruses were especially resistant to maternal plasma, suggesting that nAbs may be preventing the transmission of viral variants that are uniquely sensitive to antibody neutralization, possibly because maternal antibodies are also present in the infant at the time of transmission [45].

In addition to differences in the mode of transmission or subtype, the apparent contradictions in the susceptibility of transmitted viruses to neutralization could be explained by the timing and method of virus sampling. It remains to be determined whether there are common characteristics of all transmitted variants that might provide insights into the selective pressures that occur during the bottleneck of HIV-1 transmission. A key question in all the above studies is whether the HIV-1 present in a newly infected subject is in fact the strain that was transmitted or if it evolved shortly thereafter. Since a newly infected individual does not generate significant nAbs against HIV-1 until several months after infection, the phenotype of transmitted viruses may be coincident with replication in a non-neutralizing environment that could, for example, result in the over-production of viruses with increased neutralization sensitivity. Studies of larger panels of newly transmitted variants will be required to resolve whether they have unique properties that can be targeted by prophylactic interventions aimed at viral entry.

The synergy/antagonism of neutralizing mAbs with entry inhibitors

Several studies have investigated whether combinations of neutralizing mAbs have higher than predicted (synergistic) neutralization activity based on simple additive effects. Overall, there is no definitive answer on the possibility of synergy, beyond saying that if synergy exists it is probably quite limited [40, 46–49]. This contrasts to the strong synergy observed between CD4-IgG2 and ENF and CD4 and V3 loop-specific mAbs [50–52]. These synergies suggest that CD4 induces a structural change in the envelope trimer that increases the exposure of the ENF-binding site. There is little evidence that neutralizing mAbs induce such dramatic envelope refolding for a tangible effect on the binding of other non-overlapping mAbs [53]. Nevertheless, combinations of neutralizing mAbs, even if only additive in their effects are attractive as entry inhibitors from the standpoint of increasing the breadth of neutralization coverage, if not its potency. Of course, it is important to identify possible antagonistic effects that might blunt the prophylactic effect of a cocktail. For example, it is known that the ENF peptide and the 2F5 mAb antagonize each other, because the region in gp41 on which ENF is based includes the 2F5 epitope. A further consideration in the use of cocktails might be that they have indirect synergy, such that resistance to one inhibitor may increase sensitivity to another. For example, it has been shown that ENF resistance can increase the susceptibility of the virus to certain neutralizing mAbs, particularly those that target fusion intermediates [54]. Therefore, the use of multiple inhibitor cocktails may limit the pathways the virus has to develop resistance to any member of the cocktail.

Nonspecific inhibitors of HIV infection

A number of compounds that block HIV infection nonspecifically have been developed and tested as potential microbicides [55]. This includes surface-active agents such as cellulose acetate 1,2-benzenedicarboxylate (CAP), a polymer used to coat capsules and octylglycerol, a naturally occurring antimicrobial lipid found in human breast milk, as well as Savvy™, a surfactant. These agents coat the vagina, providing a physical barrier against infection by HIV and other pathogens. Buffering agents that restore the vagina to its naturally acidic state, such as *BufferGel*, and *Acidform* are also being tested. Cellulose sulfate, PRO 2000 and *Carraguard,* are negatively charged molecules that coat the surface of HIV, preventing it from binding to CD4 and other receptors on macrophages and dendritic cells. However, most of these compounds are fairly nonspecific for HIV and have limited potency [55]. A more specific compound is Cyanovirin-N (CVN), which is a lectin derived from blue-green algae, that binds to high mannose residues on gp120, preventing viral entry [56]. The binding site of CVN on gp120 includes the 2G12 MAb epitope [57, 58]. Preliminary data suggests that the absence of the 2G12 epitope on subtype C viruses does to compromise the ability of CVN and related compounds to inhibit HIV-1 subtype C viruses ([59] and K.B.A. unpublished data), probably because CVN binds multiple mannose residues on gp120.

Potential applications of HIV-1 entry inhibitors in developing countries

The utility of entry inhibitors for therapeutic interventions in developing countries is likely to be secondary to their use as preventative agents, given the cost and availability of these agents. There are currently a large number of clinical trials testing the efficacy of entry inhibitors as microbicides. Although first generation compounds are mainly nonspecific the newer generation of microbicides includes agents that target CCR5, such as PSC-RANTES. PSC-RANTES has been shown to protect against transmission in the macaque model and is being developed specifically for vaginal application [60]. The advantage of CCR5 antagonists is that they bind to their cellular receptors for up to 5 days and resistance takes a long time to develop, compared to other anti-retrovirals.

The finding that 2F5 and 2G12 mAbs are ineffective in neutralizing subtype C viruses *in vitro* has important implications for their use in passive immunization studies to prevent mother-child transmission in regions in Africa where subtype C circulates [40, 61]. A further hurdle to inhibiting subtype C virus entry by neutralizing mAbs is the uncertainty of the potency of 4E10 *in vivo*, as evidenced by a failure of resistance mutations to develop in a passive immunization study in which subtype B virus-infected patients that discontinued HAART were infused with mAbs to delay viral rebound [62] (Alexandra Trkola, personal communication) and the observation that this mAb is somewhat polyreactive [63]. The support for using mAbs in this context is therefore waning [64].

Conclusions

The development of compounds that block HIV-1 entry has opened up numerous possibilities for viral control, including their use as microbicides to prevent sexual transmission and to prevent transmission via breast-milk to newborns. In general, these inhibitors have so far been well tolerated and effective in limited clinical testing, and their ability to block multiple steps in the entry process makes this an attractive approach. However, the premature halting of a Phase III clinical trial of the candidate microbicides, cellulose sulfate, because of a higher number of HIV-1 infections in the active arm, is an important reminder of the limitations of in vitro testing of such compounds (http://www.who.int/mediacentre/news/statements/2007/s01/en/index.html.) Nevertheless, available data suggest that the mechanism used by HIV-1 to enter cells is the same for subtype C as for subtype B, so the same prevention strategies should be successful in subtype C infected populations, although more studies need to be done. However, irrespective of genetic subtype, HIV-l has already shown evidence that it can circumvent some of these new inhibitors either by altering the way it uses the coreceptor, by using an alternative coreceptor, or by the preexistence of polymorphisms in regions of envelope that entry inhibitors target. Thus it cannot be assumed that all entry inhibitors will be effective on HIV-1 subtype C and *in vitro* testing will be required to verify new inhibitors as they become available.

References

1 Holguin A, de Arellano R, Rivas P, Soriano V (2006) Efficacy of antiretroviral therapy in individuals infected with HIV-1 non-B subtypes. *AIDS Rev* 8: 98–107

2 Bjorndal A, Sonnerborg A, Tscherning C, Albert J, Fenyo EM (1999) Phenotypic characteristics of human immunodeficiency virus type 1 subtype C isolates of Ethiopian AIDS patients. *AIDS Res Hum Retroviruses* 15: 647–653

3 Choge I, Cilliers T, Walker P, Taylor N, Phoswa M, Meyers T, Viljoen J, Violari A, Gray G, Moore PL et al. (2006) Genotypic and phenotypic characterization of viral isolates from HIV-1 subtype C-infected children with slow and rapid disease progression. *AIDS Res Hum Retroviruses* 22: 458–465

4 Cilliers T, Nhlapo J, Coetzer M, Orlovic D, Ketas T, Olson WC, Moore JP, Trkola A, Morris L (2003) The CCR5 and CXCR4 coreceptors are both used by human immunodeficiency virus type 1 primary isolates from subtype C. *J Virol* 77: 4449–4456

5 Ping LH, Nelson JA, Hoffman IF, Schock J, Lamers SL, Goodman M, Vernazza P, Kazembe P, Maida M, Zimba D et al. (1999) Characterization of V3 sequence heterogeneity in subtype C human immunodeficiency virus type 1 isolates from Malawi: underrepresentation of X4 variants. *J Virol* 73: 6271–6281

6 Coetzer M, Cilliers T, Ping LH, Swanstrom R, Morris L (2006) Genetic characteristics of the V3 region associated with CXCR4 usage in HIV-1 subtype C isolates. *Virology* 356: 95–105

7 Connor RI, Sheridan KE, Ceradini D, Choe S, Landau NR (1997) Change in coreceptor use coreceptor use correlates with disease progression in HIV-1-infected individuals. *J Exp Med* 185: 621–628

8 Scarlatti G, Tresoldi E, Bjorndal A, Fredriksson R, Colognesi C, Deng HK, Malnati MS, Plebani A, Siccardi AG, Littman DR et al. (1997) *In vivo* evolution of HIV-1 co-receptor usage and sensitivity to chemokine-mediated suppression. *Nat Med* 3: 1259–1265

9 Clerici M, Butto S, Lukwiya M, Saresella M, Declich S, Trabattoni D, Pastori C, Piconi S, Fracasso C, Fabiani M et al. (2000) Immune activation in Africa is environmentally-driven and is associated with upregulation of CCR5. Italian-Ugandan AIDS Project. *AIDS* 14: 2083–2092

10 Pollakis G, Kang S, Kliphuis A, Chalaby MI, Goudsmit J, Paxton WA (2001) N-linked glycosylation of the HIV type-1 gp120 envelope glycoprotein as a major determinant of CCR5 and CXCR4 coreceptor utilization. *J Biol Chem* 276: 13433–13441

11 Cilliers T, Patience T, Pillay C, Papathanasopoulos M, Morris L (2004) Sensitivity of HIV type 1 subtype C isolates to the entry inhibitor T-20. *AIDS Res Hum Retroviruses* 20: 477–482

12 Trkola A, Ketas TJ, Nagashima KA, Zhao L, Cilliers T, Morris L, Moore JP, Maddon PJ, Olson WC (2001) Potent, broad-spectrum inhibition of human immunodeficiency virus type 1 by the CCR5 monoclonal antibody PRO 140. *J Virol* 75: 579–588.

13 Derdeyn CA, Decker JM, Sfakianos JN, Zhang Z, O'Brien WA, Ratner L, Shaw GM, Hunter E (2001) Sensitivity of human immunodeficiency virus type 1 to fusion inhibitors targeted to the gp41 first heptad repeat involves distinct regions of gp41 and is consistently modulated by gp120 interactions with the coreceptor. *J Virol* 75: 8605–8614

14 Reeves JD, Gallo SA, Ahmad N, Miamidian JL, Harvey PE, Sharron M, Pohlmann S, Sfakianos JN, Derdeyn CA, Blumenthal R et al. (2002) Sensitivity of HIV-1 to entry inhibitors correlates with envelope/coreceptor affinity, receptor density, and fusion kinetics. *Proc Natl Acad Sci USA* 99: 16249–16254

15 Zhang Y, Moore J (1999) Will multiple coreceptors need to be targeted by inhibitors of human immunodeficiency virus type 1 entry. *J Virol* 73: 3443–3448

16 Cilliers T, Willey S, Sullivan WM, Patience T, Pugach P, Coetzer M, Papathanasopoulos M, Moore JP, Trkola A, Clapham P et al. (2005) Use of alternate coreceptors on primary cells by two HIV-1 isolates. *Virology* 339: 136–144

17 Biscone MJ, Miamidian JL, Muchiri JM, Baik SSW, Lee FH, Doms RW, Reeves JD (2006) Functional impact of HIV coreceptor-binding site mutations. *Virology* 351: 226–236

18 Aasa-Chapman MM, Aubin K, Williams I, McKnight A (2006) Primary CCR5 only using HIV-1 isolates does not accurately represent the *in vivo* replicating quasi-species. *Virology* 351: 489–496

19 Briz V, Poveda E, Soriano V (2006) HIV entry inhibitors: mechanisms of action and resistance pathways. *J Antimicrob Chemother* 57: 619–627

20 Marozsan AJ, Kuhmann SE, Morgan T, Herrera C, Rivera-Troche E, Xu S, Baroudy BM, Strizki J, Moore JP (2005) Generation and properties of a human immunodeficiency virus type 1 isolate

resistant to the small molecule CCR5 inhibitor, SCH-417690 (SCH-D). *Virology* 338: 182–199
21 Kuhmann SE, Pugach P, Kunstman KJ, Taylor J, Stanfield RL, Snyder A, Strizki JM, Riley J, Baroudy BM, Wilson IA et al. (2004) Genetic and phenotypic analyses of human immunodeficiency virus type 1 escape from a small-molecule CCR5 inhibitor. *J Virol* 78: 2790–2807
22 Trkola A, Kuhmann SE, Strizki JM, Maxwell E, Ketas T, Morgan T, Pugach P, Xu S, Wojcik L, Tagat J et al. (2002) HIV-1 escape from a small molecule, CCR5-specific entry inhibitor does not involve CXCR4 use. *Proc Natl Acad Sci USA* 99: 395–400
23 Westby M, Lewis M, Whitcomb J, Youle M, Pozniak AL, James IT, Jenkins TM, Perros M, van der Ryst E (2006) Emergence of CXCR4-using human immunodeficiency virus type 1 (HIV-1) variants in a minority of HIV-1-infected patients following treatment with the CCR5 antagonist maraviroc is from a pretreatment CXCR4-using virus reservoir. *J Virol* 80: 4909–4920
24 Moore PL, Cilliers T, Morris L (2004) Predicted genotypic resistance to the novel entry inhibitor, BMS-378806, among HIV-1 isolates of subtypes A to G. *AIDS* 18: 2327–2330
25 Lin PF, Blair W, Wang T, Spicer T, Guo Q, Zhou N, Gong YF, Wang HG, Rose R, Yamanaka G et al. (2003) A small molecule HIV-1 inhibitor that targets the HIV-1 envelope and inhibits CD4 receptor binding. *Proc Natl Acad Sci USA* 100: 11013–11018
26 Ho HT, Fan L, Nowicka-Sans B, McAuliffe B, Li CB, Yamanaka G, Zhou N, Fang H, Dicker I, Dalterio R et al. (2006) Envelope conformational changes induced by human immunodeficiency virus type 1 attachment inhibitors prevent CD4 binding and downstream entry events. *J Virol* 80: 4017–4025
27 Cilliers T, Moore P, Coetzer M, Morris L (2005) *In vitro* generation of HIV type 1 subtype C isolates resistant to enfuvirtide. *AIDS Res Hum Retroviruses* 21: 776–783
28 Chinnadurai R, Munch J, Dittmar MT, Kirchhoff F (2005) Inhibition of HIV-1 group M and O isolates by fusion inhibitors. *AIDS* 19: 1919–1922
29 Wild CT, Shugars DC, Greenwell TK, McDanal CB, Matthews TJ (1994) Peptides corresponding to a predictive alpha-helical domain of human immunodeficiency virus type 1 gp41 are potent inhibitors of virus infection. *Proc Natl Acad Sci USA* 91: 9770–9774
30 Fouchier RA, Groenink M, Kootstra NA, Tersmette M, Huisman HG, Miedema F, Schuitemaker H (1992) Phenotype-associated sequence variation in the third variable domain of the human immunodeficiency virus type 1 gp120 molecule. *J Virol* 66: 3183–3187
31 Jensen MA, van 't Wout AB (2003) Predicting HIV-1 coreceptor usage with sequence analysis. *AIDS Rev* 5: 104–112
32 Jensen MA, Coetzer M, van 't Wout AB, Morris L, Mullins JI (2006) A reliable phenotype predictor for human immunodeficiency virus type 1 subtype C based on envelope V3 sequences. *J Virol* 80: 4698–4704
33 Baba TW, Liska V, Hofmann-Lehmann R, Vlasak J, Xu W, Ayehunie S, Cavacini LA, Posner MR, Katinger H, Stiegler G et al. (2000) Human neutralizing monoclonal antibodies of the IgG1 subtype protect against mucosal simian-human immunodeficiency virus infection. *Nat Med* 6: 200–206
34 Mascola JR (2002) Passive transfer studies to elucidate the role of antibody-mediated protection against HIV-1. *Vaccine* 20: 1922–1925
35 Mascola JR, Lewis MG, Stiegler G, Harris D, VanCott TC, Hayes D, Louder MK, Brown CR, Sapan CV, Frankel SS et al. (1999) Protection of Macaques against pathogenic simian/human immunodeficiency virus 89.6PD by passive transfer of neutralizing antibodies. *J Virol* 73: 4009–4018
36 Mascola JR, Stiegler G, VanCott TC, Katinger H, Carpenter CB, Hanson CE, Beary H, Hayes D, Frankel SS, Birx DL et al. (2000) Protection of macaques against vaginal transmission of a pathogenic HIV-1/SIV chimeric virus by passive infusion of neutralizing antibodies. *Nat Med* 6: 207–210
37 Veazey RS, Shattock RJ, Pope M, Kirijan JC, Jones J, Hu Q, Ketas T, Marx PA, Klasse PJ, Burton DR et al. (2003) Prevention of virus transmission to macaque monkeys by a vaginally applied monoclonal antibody to HIV-1 gp120. *Nat Med* 9: 343–346
38 Poignard P, Sabbe R, Picchio GR, Wang M, Gulizia RJ, Katinger H, Parren PW, Mosier DE, Burton DR (1999) Neutralizing antibodies have limited effects on the control of established HIV-1 infection *in vivo*. *Immunity* 10: 431–438
39 Binley JM, Wrin T, Korber B, Zwick MB, Wang M, Chappey C, Stiegler G, Kunert R, Zolla-Pazner S, Katinger H et al. (2004) Comprehensive cross-clade neutralization analysis of a panel of anti-human immunodeficiency virus type 1 monoclonal antibodies. *J Virol* 78: 13232–13252

40 Gray ES, Meyers T, Gray G, Montefiori DC, Morris L (2006) Insensitivity of paediatric HIV-1 subtype C viruses to broadly neutralising monoclonal antibodies raised against subtype B. *PLoS Med* 3: 1023–1031

41 Gray ES, Moore PL, Morris L (2006) Restoration of glycan 295 in subtype C viruses partially renders sensitivity to 2G12 neutralization. In: *Keystone Symposium on AIDS vaccines,* Keystone, Resort, Colorado.

42 Derdeyn CA, Decker JM, Bibollet-Ruche F, Mokili JL, Muldoon M, Denham SA, Heil ML, Kasolo F, Musonda R, Hahn BH et al. (2004) Envelope-constrained neutralization-sensitive HIV-1 after heterosexual transmission. *Science* 303: 2019–2022

43 Frost SD, Liu Y, Pond SL, Chappey C, Wrin T, Petropoulos CJ, Little SJ, Richman DD (2005) Characterization of human immunodeficiency virus type 1 (HIV-1) envelope variation and neutralizing antibody responses during transmission of HIV-1 subtype B. *J Virol* 79: 6523–6527

44 Chohan B, Lang D, Sagar M, Korber B, Lavreys L, Richardson B, Overbaugh J (2005) Selection for human immunodeficiency virus type 1 envelope glycosylation variants with shorter V1–V2 loop sequences occurs during transmission of certain genetic subtypes and may impact viral RNA levels. *J Virol* 79: 6528–6531

45 Wu X, Parast AB, Richardson BA, Nduati R, John-Stewart G, Mbori-Ngacha D, Rainwater SM, Overbaugh J (2006) Neutralization escape variants of human immunodeficiency virus type 1 are transmitted from mother to infant. *J Virol* 80: 835–844

46 Mascola JR, Louder MK, VanCott TC, Sapan CV, Lambert JS, Muenz LR, Bunow B, Birx DL, Robb ML (1997) Potent and synergistic neutralization of human immunodeficiency virus (HIV) type 1 primary isolates by hyperimmune anti-HIV immunoglobulin combined with monoclonal antibodies 2F5 and 2G12. *J Virol* 71: 7198–7206

47 Verrier F, Nadas A, Gorny MK, Zolla-Pazner S (2001) Additive effects characterize the interaction of antibodies involved in neutralization of the primary dualtropic human immunodeficiency virus type 1 isolate 89.6. *J Virol* 75: 9177–9186

48 Xu W, Smith-Franklin BA, Li PL, Wood C, He J, Du Q, Bhat GJ, Kankasa C, Katinger H, Cavacini LA et al. (2001) Potent neutralization of primary human immunodeficiency virus clade C isolates with a synergistic combination of human monoclonal antibodies raised against clade B. *J Hum Virol* 4: 55–61

49 Zwick MB, Wang M, Poignard P, Stiegler G, Katinger H, Burton DR, Parren PW (2001) Neutralization synergy of human immunodeficiency virus type 1 primary isolates by cocktails of broadly neutralizing antibodies. *J Virol* 75: 12198–12208

50 Nagashima KA, Thompson DA, Rosenfield SI, Maddon PJ, Dragic T, Olson WC (2001) Human immunodeficiency virus type 1 entry inhibitors PRO 542 and T-20 are potently synergistic in blocking virus-cell and cell-cell fusion. *J Infect Dis* 183: 1121–1125

51 Crooks ET, Moore PL, Richman D, Robinson J, Crooks JA, Franti M, Schulke N, Binley JM (2005) Characterizing anti-HIV monoclonal antibodies and immune sera by defining the mechanism of neutralization. *Hum Antibodies* 14: 101–113

52 Potts BJ, Field KG, Wu Y, Posner M, Cavacini L, White-Scharf M (1993) Synergistic inhibition of HIV-1 by CD4 binding domain reagents and V3-directed monoclonal antibodies. *Virology* 197: 415–419

53 Kwong PD, Doyle ML, Casper DJ, Cicala C, Leavitt SA, Majeed S, Steenbeke TD, Venturi M, Chaiken I, Fung M et al. (2002) HIV-1 evades antibody-mediated neutralization through conformational masking of receptor-binding sites. *Nature* 420: 678–682

54 Reeves JD, Lee FH, Miamidian JL, Jabara CB, Juntilla MM, Doms RW (2005) Enfuvirtide resistance mutations: impact on human immunodeficiency virus envelope function, entry inhibitor sensitivity, and virus neutralization. *J Virol* 79: 4991–4999

55 Hillier S (2006) Microbicides: State of the Art and Its Evolution. In: *Microbicides 2006 Conference*: Cape Town.

56 Boyd MR, Gustafson KR, McMahon JB, Shoemaker RH, O'Keefe BR, Mori T, Gulakowski RJ, Wu L, Rivera MI, Laurencot CM et al. (1997) Discovery of cyanovirin-N, a novel human immunodeficiency virus-inactivating protein that binds viral surface envelope glycoprotein gp120: potential applications to microbicide development. *Antimicrob Agents Chemother* 41: 1521–1530

57 Esser MT, Mori T, Mondor I, Sattentau QJ, Dey B, Berger EA, Boyd MR, Lifson JD (1999) Cyanovirin-N binds to gp120 to interfere with CD4-dependent human immunodeficiency virus type 1 virion binding, fusion, and infectivity but does not affect the CD4 binding site on gp120 or soluble CD4-induced conformational changes in gp120. *J Virol* 73: 4360–4371

58 Witvrouw M, Fikkert V, Hantson A, Pannecouque C, O'Keefe B R, McMahon J, Stamatatos L, de Clercq E, Bolmstedt A (2005) Resistance of human immunodeficiency virus type 1 to the high-mannose binding agents cyanovirin N and concanavalin A. *J Virol* 79: 7777–7784

59 Alexandre K, Gray E, Moore P, Taylor N, Chikwamba R, O'Keefe B, Morris L (2006) Sensitivity of HIV-1 subtype C viruses to Griffithsin and Scytovirin: potential HIV microbicides In: *Microbicides 2006 Conference*: Cape Town.

60 Lederman MM, Veazey RS, Offord R, Mosier DE, Dufour J, Mefford M, Piatak M Jr, Lifson JD, Salkowitz JR, Rodriguez B et al. (2004) Prevention of vaginal SHIV transmission in rhesus macaques through inhibition of CCR5. *Science* 306: 485–487

61 Safrit JT, Ruprecht R, Ferrantelli F, Xu W, Kitabwalla M, Van Rompay K, Marthas M, Haigwood N, Mascola JR, Luzuriaga K et al. (2004) Immunoprophylaxis to prevent mother-to-child transmission of HIV-1. *J Acquir Immune Defic Syndr* 35: 169–177

62 Trkola A, Kuster H, Rusert P, Joos B, Fischer M, Leemann C, Manrique A, Huber M, Rehr M, Oxenius A et al. (2005) Delay of HIV-1 rebound after cessation of antiretroviral therapy through passive transfer of human neutralizing antibodies. *Nat Med* 11: 615–622

63 Haynes BF, Fleming J, St Clair EW, Katinger H, Stiegler G, Kunert R, Robinson J, Scearce RM, Plonk K, Staats HF et al. (2005) Cardiolipin polyspecific autoreactivity in two broadly neutralizing HIV-1 antibodies. *Science* 308: 1906–1908

64 Gorny MK, Zolla-Pazner S (2006) Immunoprophylaxis against mother-to-child transmission of HIV-1. *PLoS Med* 3: 966–967

65 Bures R, Morris L, Williamson C, Ramjee G, Deers M, Fiscus SA, Abdool-Karim S, Montefiori DC (2002) Regional clustering of shared neutralization determinants on primary isolates of clade C human immunodeficiency virus type 1 from South Africa. *J Virol* 76: 2233–2244

Entry Inhibitors in HIV Therapy
Edited by Jacqueline D. Reeves and Cynthia A. Derdeyn

The utility of coreceptor typing in the clinic

Eoin Coakley

Monogram Biosciences, 345 Oyster Point Blvd, South San Francisco, CA 94080, USA

Introduction

HIV enters cells through critical interactions of the viral envelope protein (Env) with the CD4 receptor and a chemokine coreceptor, typically CCR5 or CXCR4 [1]. This chapter focuses on these two most clinically relevant coreceptors. Based upon the pattern of coreceptor utilization, a virus may be classified as exhibiting one of three phenotypes; a virus using only CCR5 is R5 tropic, one that uses only CXCR4 is X4 and one that can use either coreceptor is R5X4 or dual tropic. Since HIV commonly exists within a patient as a mixture of viruses with different tropisms, there is a fourth possible determination, mixed tropism. For example, a single HIV plasma sample may contain R5 as well as X4 or R5X4 viruses. Population-based coreceptor typing assays cannot distinguish between truly dual tropic viral populations and those that are mixed. Therefore the detection of both R5 and X4 tropism in a sample is reported as dual/mixed (DM).

In the advent of coreceptor inhibitors, clinical trials have incorporated coreceptor typing at patient screening, study entry and/or during treatment with coreceptor inhibitors for monitoring purposes, as CCR5 inhibitors are unlikely to be effective against X4 viruses and *vice versa*. Furthermore, given the link between CXCR4 use and disease progression, monitoring the effect of CCR5 inhibitors on viral tropism and clinical outcome becomes imperative.

Determination of coreceptor tropism

MT-2 cell assay

Prior to the identification of HIV coreceptors, HIV strains were often characterized as syncytium inducing (SI) or non-syncytium inducting (NSI) based upon their ability to infect and mediate multinucleated giant cell (syncytia) formation in MT-2 cells [2]. Subsequently, the ability of a strain of HIV to infect and cause syncytia formation in MT-2 was found to correlate with utilization of the CXCR4 coreceptor as MT-2 cells express CD4 and CXCR4 but not the CCR5 coreceptor. Broadly, two standardized MT-2 assay approaches are utilized. In one there is a requirement to generate viral stocks from stimulated patient lymphocytes co-cultured with lymphocytes from an HIV-negative donor [3]. These stocks are titrated and then used to infect MT-2 cells. The time needed from infection to assay readout is typically 14 days or longer. Readout requires microscopic inspection of individual cultures to determine the presence (SI) or absence (NSI) of syncytia. A concurrent infection of donor lymphocytes serves as a control culture to confirm viability of the patient HIV stock. The second method utilizes direct co-cultivation of patient lymphocytes with MT-2 cells, followed by microscopic examination [2]. Up to recently, the MT-2 assay was a common method of determining HIV phenotype in clinical research settings. Indeed, early studies utilizing the MT-2 assay established an SI phenotype as one of the first markers of disease progression [4]. Despite these findings, the MT-2 assay did not become a routine clinical monitoring test in the way that plasma HIV RNA quantitation and CD4 cell counts have. There are a number of possible reasons for this; (1) the time- and labor-dependent nature of the assay process, (2) the lack of ability to directly alter this phenotype by previously available antiretrovirals, (3) the virus tested is derived from stimulated lymphocytes and not plasma virus and may thus not be representative of circulating virus, and (4) the non-quantitative nature of the assay readout as SI or NSI.

Recombinant virus phenotypic assays

Available commercial coreceptor tropism assays [5–7] involve the amplification of *env* genes from patient plasma samples. The resulting amplicons are used to generate recombinant HIV constructs, or are cloned into an Env expression vector and used to pseudotype a recombinant HIV reporter vector. Recombinant viruses or reporter pseudotypes are then used to infect mammalian cell lines expressing CD4 and either CXCR4 or CCR5. One such high-throughput assay (Trofile™, Monogram Biosciences) has been incorporated into current clinical trial protocols for coreceptor inhibitor testing. This single-cycle assay utilizes luciferase reporter pseudotype viruses and quantitates luciferase activity as relative light units (RLUs) to define infection of U87 cells expressing CD4 and either CXCR4 or CCR5. As a confirmatory step, luciferase production must also be able to be inhibited by an antagonist specific for the coreceptor being tested. This step is particularly relevant when infection levels are low, resulting in luciferase activity close to background levels.

CXCR4 utilization is common among clinical isolates from individuals with advanced disease and, as dual or X4 viruses might naturally evade the effects of CCR5 inhibitors, there is great interest in the sensitivity of current assays to detect these X4 viruses. Studies of the ability of recombinant virus assays to detect minor species of X4 and/or R5 viruses show limits of detection in the range of 5–10% [7]. It is not currently known if X4 or dual viruses at very low levels impact the activity of a CCR5 antagonist when used in combination anti-HIV therapy. It is possible that detection of any CXCR4 activity might negatively impact outcome. Alternatively low-level CXCR4 use may have no meaningful impact on treatment outcome. It is anticipated that these important questions will be addressed as the current Phase III tropism antagonist trial data emerge.

Comparison of MT-2 and recombinant virus coreceptor tropism assays

There are important differences between MT-2 and recombinant virus assays. These assays typically evaluate HIV from distinct compartments; stimulated lymphocytes *versus* plasma. The MT-2 assay utilizes native virus and recombinant assays evaluate the complete viral Env or a fragment of Env [5–7]. The MT-2 assay permits multiple cycles of replication (and possible amplification of viral subpopulations and/or viral adaptation to culture conditions), while recombinant assays limit replication to a single cycle.

An SI result in an MT-2 assay is an established surrogate for CXCR4 utilization. Currently, only very limited data are available examining the relationship between phenotypes determined by the MT-2 and the Trofile coreceptor tropism assay. However, in one study, 11 individuals with HIV determined to be SI in the MT-2 assay [8] has coreceptor typing performed retrospectively with Trofile and all 11 isolates showed X4 or DM tropism. Interestingly, the CXCR4 luciferase activity among these 11 SI isolates was not uniform but rather varied over a very broad range of luciferase activity. Further studies will be required to determine whether this is clinically meaningful.

In a second study, the AIDS Clinical Trials Group 5211 study, the Trofile assay was utilized to coreceptor type individuals prior to entry into a clinical trial of vicriviroc [9]. MT-2 assays were performed retrospectively among baseline isolates and demonstrated only limited discordance between the two assays [10]. Equally importantly in this study perhaps was the low (50%) HIV recovery rate among lymphocyte samples processed for the MT-2 assay compared to samples successfully phenotype by the Trofile assay.

Genotyping of the V3 loop

The genetic determinants of coreceptor tropism reside largely, but not exclusively, in the V3 region of gp120. A variety of algorithms have been developed to approximate viral phenotype/coreceptor tropism (NSI/SI or X4, R5) using sequence data from this region. These include the 11/25 charge rule, decision tree analysis, the position-specific scoring matrix (PSSM) and the support vector machine (SVM) [11–14]. The earliest of these, the 11/25 charge rule, was based on the observation that SI viruses often possess positively charged (basic) amino acids at codons 11 and/or 25 in the V3 loop. These observations have since been enlarged upon and it is now appreciated that changes at other codons may also influence phenotype. More recent algorithms, SVM and PSSM, evaluate more V3 loop sequence information and are regarded as being advances on the original 11/25 charge rule. However, there are several significant challenges for the use of algorithms in predicting coreceptor tropism from sequences derived from a patients viral population [15]. These include: (1) Env sequence data derived from viral populations is more challenging to interpret than that from clonal viruses since mixed populations of viruses with different V3 loop sequences can co-exist and because of the difficulty of Env sequence alignment due to extensive Env variability and length polymorphisms. (2) Current algorithms are optimized for subtype B viruses, but are not yet fully defined for non-B subtypes. (3) Sequencing does not distinguish between genetically viable and non-viable Envs, and (4) sequences outside of V3 can be important determinants of tropism [16].

Coreceptor tropism epidemiology

Recombinant virus assay data

Several groups have evaluated plasma HIV coreceptor tropism patterns using recombinant viruses in treated and untreated populations (Tab. 1).

Treatment-naïve individuals

Harrigan and colleagues [17] evaluated tropism profiles in 979 antiretroviral naïve individuals in the HAART Observational Medical Evaluation and Research (HOMER) study in British Columbia, Canada. The median CD4 count in this population was 260 cells/mm^3. The overall prevalence of samples demonstrating DM or X4 tropism was 18%, while 82% were R5. In this study the prevalence of any X4 tropism (X4 or DM) increased progressively with CD4 cell counts below 200/mm^3. In fact, detection of any X4 variants was 16.6-fold more frequent in those with CD4 counts less than 25/mm^3 compared with those with counts above 500/mm^3.

Treatment-experienced individuals

The prevalence of X4 tropic viruses in highly treatment-experienced individuals has been evaluated in several studies [18–20]. The AIDS Clinical Trials Group Study 5211 evaluated an antiretroviral-experienced population with virological failure of a ritonavir-boosted protease inhibitor regimen (Tab. 1). Among those screening for this study, the median CD4 cell count was

Table 1. Prevalence of R5 compared to X4, dual/mixed (DM) profiles in various treated and untreated HIV positive populations[a]

	HOMER [17]	Demarest [19]	HGDS [29]	Demarest [19]	SCOPE [21]	TORO31 [20]	ACTG 5211 [18]
Antiretroviral naive	Yes	Yes	Yes	No	No	No	No
X4 or DM[b]	18%	12%	40%	33%	40%	50%	51%
R5	82%	88%	60%	67%	60%	50%	49%
CD4 (median)	260	-	409	-	258	80/93[c]	140
n	979	325	125	117	186	627	321

[a] A variety of different populations are represented here. The HOMER cohort from British Colombia defined tropism profiles in antiretroviral naive subjects. Demarest et al. [19] evaluated tropism in both antiretroviral naïve and experienced clinic populations. The HGDS (Hemophilia Growth and Development Study) looked at tropism profiles in pediatric/adolescent hemophiliacs who were largely antiretroviral naïve (50% were receiving treatment with a single nucleoside analog). The SCOPE study population comprised a population of patients who were experiencing ongoing failure of potent antiretroviral therapy and who were clinically stable. The TORO trials were multinational registrational trials of the HIV fusion inhibitor enfuvirtide. These individuals were more advanced in disease stage than the other studies described here. Data from the screening phase of ACTG 5211 study are shown. This was a Phase II study of the safety and efficacy of vicriviroc in treatment-experienced subjects. This population is similar to that evaluated in the TORO trials.
[b] X4 (CXCR4 only tropism); DM, dual or mixed tropism; R5, CCR5 only tropism.
[c] Median CD4 counts for TORO 1 and TORO 2 trials, respectively.

140/mm^3 [18]. The prevalence of DM or X4 virus was 51%, which is higher than that seen in the various antiretroviral-naïve cohorts. The observed higher prevalence of X4 tropism in treatment-experienced individuals was further explored by Hunt and colleagues within the SCOPE (Study of the Consequences of the Protease inhibitor Era) cohort [21]. This cohort focuses on individuals with ongoing viremia in the setting of stable HIV therapy. The prevalence of CXCR4 utilizing virus was compared to that of antiretroviral naïve individuals within the HOMER cohort referred to above. At each CD4 quartile the prevalence of CXCR4 utilization was significantly higher in those with treatment failure compared to untreated individuals. Recent data suggest that this phenomenon may also be seen in clade C viruses [22]. It is not currently know why there might be a relative excess of X4 tropism in antiretroviral-experienced individuals with treatment failure stratified by CD4 count. However, this important observation has relevance to the future design of trials with coreceptor antagonists and to the potential utilization of CCR5 antagonists in the clinic. For example, if X4 detection limits the antiviral activity of CCR5 antagonists *in vivo* then a smaller proportion of treatment-experienced population would be anticipated to benefit from the antiviral activity of this new class compared to treatment-naïve populations.

Coreceptor tropism and disease progression

Using an MT-2 cell assay, Koot et al. [4] made the original observation within the Amsterdam cohort that antiretroviral-naive individuals with SI virus experienced more rapid disease progression compared to individuals with NSI virus. This observation has subsequently been confirmed in other cohorts of HIV-positive individuals [23–28]. A recombinant virus assay was utilized to assess the impact of plasma HIV coreceptor tropism on disease progression by Daar and colleagues [29] in the Hemophilia Growth and Development Study. This was a retrospective study of hemophiliacs infected at a time when highly active antiretroviral therapy was not generally available. Coreceptor usage was defined in 95 subjects. At entry, 49 of 95 subjects had received prior nucleoside reverse transcriptase inhibitor monotherapy. To a maximum follow-up of 8 years, the rate of progression to AIDS was significantly greater for those with any X4 tropism. In multivariate models including baseline CD4 count, HIV-1 RNA and antiviral therapy, any X4 tropism was an independent predictor of progression to AIDS.

Tropism switches

It was observed relatively early in the HIV epidemic that coreceptor tropism was not fixed at primary infection but could evolve over time. After primary HIV infection, most viruses are NSI with evolution to SI occurring later in some but not all individuals. Data from the Amsterdam Cohort demonstrated that a 'switch' in phenotype from NSI to SI phenotype was temporally associated with accelerated CD4 decline [4]. Clinicians may assume that the so-called 'switch' from NSI to SI phenotype implies a switch from exclusive CCR5 to exclusive CXCR4 utilization. In fact, more commonly this 'switch' represents the emergence of isolates that are able to use of both coreceptors, i.e., dual tropism [30]. Similar observations have also been made utilizing plasma HIV tested by a recombinant virus assay [31].

Tropism in the era of coreceptor inhibitors

One of the most significant recent developments in HIV therapy has been the targeting of steps in the HIV entry process: CD4 attachment, coreceptor attachment, and fusion. One such agent, enfuvirtide (Fuzeon, T-20), a fusion inhibitor, is approved for use in the clinic. Drugs that target either CXCR4 or CCR5 coreceptors are currently in various stages of clinical development (Tab. 2). One reason blockade of human coreceptors might serve as a clinically useful anti-HIV target is because individuals who are deficient in or who lack cell surface CCR5 expression are relatively common in some populations, and these individuals appear to be healthy. The *ccr5Δ32* mutation results in a

Table 2. Clinical trials of coreceptor tropism antagonists

Agent	Trial name	Antiretroviral experience	Screening tropism	Study anti-HIV regimens
Maraviroc	A4001026[a]	Naïve	R5	EFV or maraviroc once daily or twice daily each with 3TC-ZDV
	A4001027	Experienced (North America)	R5	OB Vs OB + maraviroc
	A4001028	Experienced (Global)	R5	OB Vs OB + maraviroc
	A4001029	Experienced	DM	OB Vs OB + maraviroc
Vicriviroc	P03820[b]	Naïve	R5	3TC-ZDV + EFV or vicriviroc
	ACTG 5211	Experienced	R5	vicriviroc + OB
Aplaviroc[c]	CCR-100136	Naïve	R5	LPV/r + aplaviroc
	CCR-102881	Naïve	R5	ZDV-3TC + aplaviroc or EFV
AMD11070	ACTG 5210	N/A[d]	X4	11070 monotherapy
	XACT	N/A[d]	X4	11070 monotherapy

[a] The arm comprising maraviroc once daily with ZDV and 3TC was terminated early to due inferior virological outcomes compared to the reference group of EFV with ZDV and 3TC.
[b] This trial was closed due to inferior antiviral activity in the vicriviroc group.
[c] Aplaviroc is no longer in clinical development due to the observation of possible drug-associated hepatitis in clinical trials.
[d] In AMD 11070 trials antiretroviral experienced and naïve individuals were able to enroll but were required to be on no antiretrovirals at the time of study.
3TC: lamivudine, ZDV: zidovudine, EFV: efavirenz, OB: optimized background, fixed dose LPV/r lopinavir/ritonavir.

naturally occurring phenotype in which fully functional CCR5 coreceptors are either absent (*ccr5Δ32* homozygous) or present at reduced levels (*ccr5Δ32* heterozygous) [32]. Individuals homozygous or heterozygous for this deletion appear to be otherwise healthy without significant perturbation of their immune function. Indeed, those homozygous for the *ccr5Δ32* mutation appear to be protected from infection by R5 HIV. Conversely, such individuals may be at higher risk for West Nile virus encephalitis [33]. It remains to be fully defined why this phenotype is particularly common in those of Northern European decent relative to other ethnic/geographic groups. Nevertheless, the existence of apparently healthy individuals who are deficient in or who lack CCR5 expression is relevant to the investigation of coreceptor antagonism as a target for anti-HIV therapies and several antagonists have entered clinical trials. It should be noted that antagonists of the CXCR4 coreceptor are also in clinical development as possible anti-HIV therapeutics [34].

A key issue for this new class is that each drug under investigation blocks utilization of either coreceptor (most target CCR5) but not both. It is plausible that CCR5 blockade may fail to fully suppress HIV replication if CXCR4 utilizing viruses are also present. In this setting it is also possible that exposure to such a drug might shift a mixed tropic predominantly R5 population to predominantly X4. This concept received some validation from Pfizer's clinical trial with maraviroc (MVC) in protocol A4001029 [35]. This was a Phase IIb pilot study, which examined the impact of exposure to MVC, a CCR5 antagonist, in individuals with dual or mixed tropic virus. The 167 individuals were treatment-experienced and received concurrent optimized background therapy (OBT). At week 24 there was no difference in HIV RNA outcomes in the group receiving MVC with OBT compared those receiving OBT only. As anticipated, in some individuals a shift from DM to X4-only tropism was observed in the setting of MVC exposure. Intriguingly, the shift to X4-only tropism was not associated with CD4 declines during limited follow-up. These data support the concept that the detection of CXCR4 utilizing virus at screening will limit the antiviral activity of CCR5 inhibitors and are supportive of the concept of coreceptor screening prior to use of this drug class.

The degree to which X4 or dual tropic virus exist at some level in an individual determined to have R5 virus by population-based methods is not known. It has been shown that X4 viruses may exist at levels below the limits of detection by current assays [30]. Monotherapy studies of CCR5 antagonists have provided some insight in to this possibility [36, 37]. Investigators evaluated monotherapy with the CCR5 antagonist aplaviroc (APL) in antiretroviral-naïve individuals with R5-only virus at baseline. Of the 31 patients receiving 10 days of APL monotherapy, onc individual had detectable CXCR4-using virus at study end. Analysis of viable *env* clones demonstrated that 4% were actually CXCR4 using at baseline and at day 10 this had risen to 27%. At day 24 (14 days post APL exposure), the proportion had fallen to 8%. It is also reasonable to speculate that dual or X4 variants that exist in reservoirs and that are not represented in the periphery will be "invisible" in assays evaluating the coreceptor tropism of circulating viruses. It is anticipated that current clinical trials will provide some insight into the prevalence of low-level circulating X4 variants and their possible relevance to clinical outcomes. There are a number of possible outcomes with regard to the emergence of X4 viruses among individuals using coreceptor antagonists as components of antiretroviral therapy (Fig. 1). Firstly, individuals with R5 virus may experience sustained suppression of HIV RNA levels on therapy. Secondly, some individuals with R5 virus may have X4 virus present at levels below the limits of detection. In such cases, suppression of the R5 component may 'unmask' underlying X4 variants, which may continue to replicate at low levels (low X4 viremia) or alternatively, may replicate to higher levels (high X4 viremia). Finally, it is possible that individuals with R5 viruses may experience a late viral load rebound after a period of sustained viral suppression. This may reflect a tropism switch and

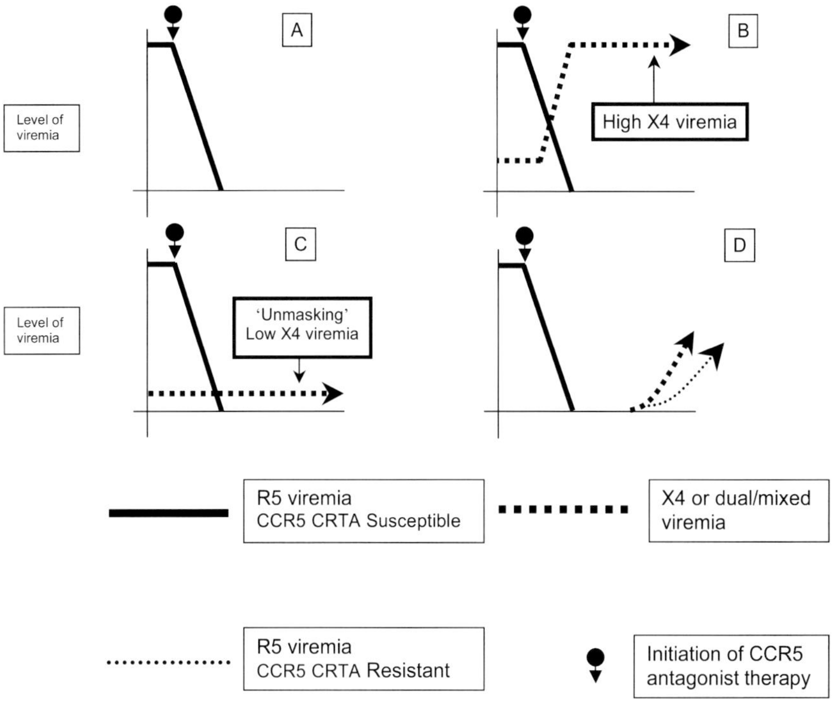

Figure 1. Theoretical viremia outcomes with CCR5 antagonist therapy. (A) R5 viremia suppressed with combination therapy including a CCR5 antagonist. (B) X4 viremia exists at levels low levels. On suppression of R5 viremia, X4 viremia emerges at high level (High X4 Viremia). (C) X4 viremia exists at low levels (possibly below the limits of detection). On suppression of R5 viremia, X4 viremia remains at a sustained low level (Low X4 Viremia) possibly being "unmasked". (D) R5 viremia suppressed with combination therapy including a CCR5 antagonist but with late emergence of a population that is X4 or phenotypically resistant to a CCR5 antagonist. CRTA: coreceptor tropism antagonist.

acquisition of X4 variants and/or selection of virus that remain R5 tropic but have become drug resistant.

Finally, it is not clear whether the correlation between SI (or X4, DM) phenotype and disease progression, which has been documented in the natural history studies described above, will be as relevant in the setting of failure of potent combination antiretroviral therapy and more particularly in the setting of failure of coreceptor antagonists. It is possible that the X4 or DM virus emerging in these setting may not be associated with either accelerated disease progression or CD4 decline. This is in part because it is not yet established the degree to which the emergence of CXCR4 utilizing strains is a consequence or a cause of the altered host immune environment. It is anticipated that these and the other issues raised in this discussion may be answered more fully in the ongoing Phase III clinic trials.

Summary

The emergence of coreceptor antagonists as a possible new class of anti-HIV drugs is a reason for further optimism in the treatment of HIV. A feature of this new drug class, however, is the possibility that circulating HIV may posses a coreceptor tropism profile that might allow the virus to evade the effects of these new drugs. This highlights the likelihood that coreceptor tropism determination will be relevant to the optimal use of this new drug class, as a screening tool prior to drug use and also possibly in the setting of treatment failure with a coreceptor antagonist. It is anticipated that the current clinical trials of these agents will provide further insight into the utility of coreceptor typing.

Acknowledgements
The author would like to acknowledge Christos Petropoulos, Jeannette Whitcomb, Wei Huang, Eric Stawiski and Jackie Reeves for their assistance in the preparation of this chapter.

References

1 Doranz BJ, Berson JF, Rucker J, Doms RW (1997) Chemokine receptors as fusion cofactors for human immunodeficiency virus (HIV-1). *Immunol Res* 16: 15–28
2 Koot M, Vos AHV, Keet RPM, de Goede RE, Dercksen MW, Terpstra FG, Coutinho RA, Miedema F, Tersmette M (1992) HIV-1 biological phenotype in long term infected individuals, evaluated with an MT-2 cocultivation assay. *AIDS* 6: 49–54
3 The ACTG Virology Technical Advisory Committee and the Division of AIDS, National Institutes of Health (1997) *The ACTG Virology Manual for HIV laboratories*. NIH, Bethesda 1997
4 Koot M, Keet IP, Vos AH, de Goede RE, Roos MT, Coutinho RA, Miedema F, Schellekens PT, Tersmette M (1993) Prognostic value of HIV-1 syncytium-inducing phenotype for rate of $CD4^+$ cell depletion and progression to AIDS. *Ann Intern Med* 118: 681–688
5 Trouplin V, Salvatori F, Cappello F, Obry V, Brelot A, Heverker N, Alizon M, Scarlatti G, Clavel F, Mammano F (2001) Determination of coreceptor usage of human immunodeficiency virus type 1 from patient plasma samples by using a recombinant phenotypic assay. *J Virol* 75: 251–259
6 Huang W, Wrin M, Yap J, Fransen S, Beauchaine T, Reddy M, Paxinos EE, Parkin NT, Whitcomb JM, Petropoulos CJ (2001) A rapid multi-functional HIV-1 entry assay for measuring drug susceptibility, coreceptor tropism, and antibody neutralization. 41st Interscience Conference on Antimicrobial Agents and Chemotherapeutics; December 2001, Chicago, IL, abstr. no. I-1968
7 Whitcomb JM, Huang W, Fransen S, Limoli K, Toma J, Wrin T, Chappey C, Kiss L, Paxinos EE, Petropoulos CJ (2007) Development and characterization of a novel single-cycle recombinant virus assay to determine human immunodeficiency virus type-1 co-receptor tropism. *Antimicrob Agents Chemother*. 51: 566–575
8 The AMD3100 HIV Study Group (2004) Safety, pharmacokinetics, and antiviral activity of AMD3100, a selective CXCR4 receptor inhibitor, in HIV-1 infection. *J Acquir Immune Defic Syndr* 37: 1253–1262
9 Gulick R, Su Z., Flexner C et al., for the ACTG 5211 Study Team (2006) ACTG 5211: Phase II study of the safety and efficacy of vicriviroc in HIV-infected treatment-experienced subjects. Program and abstracts of the XVI International AIDS Conference; August 13–18, 2006, Toronto, Ontario, abstr. no. THLB0217
10 Hosoya N, Su Z, Wilkin T, Gulick R, Flexner C, Hughes M, Skolnik P, Greaves W, Coakley E, Kuritzkes D (2006) Assessing HIV-1 tropism in ACTG 5211: A comparison of assays using replication competent virus from peripheral blood mononuclear cells *versus* plasma-derived pseudotyped virions. Abstracts of the 2nd International Workshop Targeting HIV Entry; October 20–21, 2006, Boston, abstr. no. 33
11 Hoffman NG, Seillier-Moiseiwitsch F, Ahn J, Walker JM, Swanstrom R (2002) Variability in the

human immunodeficiency virus type 1 gp120 Env protein linked to phenotype-associated changes in the V3 loop. *J Virol* 76: 3852–3864
12 Pillai S, Good B, Richman D, Corbeil J (2003) A new perspective on V3 phenotype prediction. *AIDS Res Hum Retroviruses* 19: 145–149
13 Jensen MA, Li FS, van 't Wout AB, Nickle DC et al. (2003) Improved coreceptor usage prediction and genotypic monitoring of R5-to-X4 transition by motif analysis of human immunodeficiency virus type 1 env V3 loop sequences. *J Virol* 77: 13376–13388
14 Sing T, Beerenwinkel N, Lengauer T (2004) Learning mixtures of localized rules by maximizing the area under the ROC curve. 16th European Conference on Artificial Intelligence (ECAI), Workshop on ROC Analysis in AI , 2004
15 Stawiski E, Liu Y, Toma J, Huang W, Fransen S, Whitcomb J, Coakley E, Parkin N, Eshleman S, Petropoulos C, Chappey C (2006) Limitations in predicting HIV-1 co-receptor tropism from V3 genotype data. XVI International AIDS Conference, Toronto, Canada, August 13–18, 2006, abstr. no. THPE0046
16 Huang W, Eshelman S, Toma J, Fransen S, Stawiski E, Jackson B, Whitcomb J, Parkin N, Petropoulos C (2006) CXCR4 use is common in subtype D HIV-1; Co-receptor tropism determinants exist outside of the V3 region of gp120. Abstracts of the 2nd International Workshop Targeting HIV Entry; October 20–21, 2006, Boston, abstr. no. 7
17 Brumme ZL, Goodrich J, Mayer HB, Brumme CJ, Henrick BM, Wynhoven B, Asselin JJ, Cheung PK, Hogg RS, Montaner JS, Harrigan PR (2005) Molecular and clinical epidemiology of CXCR4-using HIV-1 in a large population of antiretroviral-naive individuals. *J Infect Dis* 192: 466–474
18 Wilkin T, Su Z, Kuritzkes D, Hughes M, Flexner C, Gross R, Coakley E, Greaves W, Godfrey C, Skolnik PR (2006) Co-receptor tropism in patients screening for ACTG 5211, a phase 2 study of vicriviroc, a CCR5 inhibitor. *Clin Infect Disease* 44: 591–595
19 Demarest J, Bonny T, Vavro C, LaBranche C, Kitrinos K, McDanal C, Sparks S, Chavers S, Castillo S, Elrick D (2004) HIV-1 co-receptor tropism in treatment naive and experienced subjects. Abstracts of the 44th Interscience Conference on Antimicrobial Agents and Chemotherapy (ICAAC), October 30–November 2, 2004, Washington, DC, abstr. no. H-1136
20 Melby T, Despirito M, Demasi R, Greenberg M (2006) HIV-1 Co-receptor tropism in triple-class-experienced patients: baseline correlates and relationship to enfuvirtide response. Abstracts of the 13th Conference on Retroviruses and Opportunistic Infections, Denver, February 5–9, 2006, abstr. no. 233
21 Hunt PW, Harrigan PR, Huang W (2006) Prevalence of CXCR4 tropism among antiretroviral-treated HIV-1-infected patients with detectable viremia. *J Infect Dis* 194: 926–930
22 Johnston ER, Zijenah LS, Mutetwa S, Kantor R, Kittinunvorakoon C, Katzenstein DA (2003) High frequency of syncytium-inducing and CXCR4-tropic viruses among human immunodeficiency virus type 1 subtype C-infected patients receiving antiretroviral treatment. *J Virol* 77: 7682–7688
23 Bozzette SA, McCutchan JA, Spector SA, Wright B, Richman DD (1993) A cross-sectional comparison of persons with syncytium- and non-syncytium-inducing human immunodeficiency virus. *J Infect Dis* 168: 1374–1379
24 Hughes MD, Johnson VA, Hirsch MS, Bremer JW, Elbeik T, Erice A, Kurtizkes DR, Scott WA, Spector SA, Basgoz N et al. (1997) Monitoring plasma HIV-1 RNA levels in addition to $CD4^+$ lymphocyte count improves assessment of antiretroviral therapeutic response. ACTG 241 Protocol Virology Substudy Team. *Ann Intern Med* 126: 929–938
25 Japour AJ, Welles S, D'Aquila RT, Johnson VA, Richman DD, Coombs RW, Reichelderfer PS, Kahn JO, Crumpacker CS, Kuritzkes DR (1995) Prevalence and clinical significance of zidovudine resistance mutations in human immunodeficiency virus isolated from patients after long-term zidovudine treatment. AIDS Clinical Trials Group 116B/117 Study Team and the Virology Committee Resistance Working Group. *J Infect Dis* 171: 1172–1179
26 Lathey JL, Hughes MD, Fiscus SA, Pi T, Jackson JB, Rasheed S, Elbeik T, Reichman R, Japour A, D'Aquila RT et al. (1998) Variability and prognostic values of virologic and CD4 cell measures in human immunodeficiency virus type 1-infected patients with 200–500 CD4 cells/mm^3 (ACTG 175). AIDS Clinical Trials Group Protocol 175 Team. *J Infect Dis* 177: 617–624
27 Sheppard HW, Lang W, Ascher MS, Vittinghoff E, Winkelstein W (1993) The characterization of non-progressors: long-term HIV-1 infection with stable $CD4^+$ T-cell levels. *AIDS* 7: 1159–1166
28 Spijkerman I, de Wolf F, Langendam M, Schuitemaker K, Coutinho R (1998) Emergence of syncytium-inducing human immunodeficiency virus type 1 variants coincides with a transient

increase in viral RNA level and is an independent predictor for progression to AIDS. *J Infect D*is 178: 397–403

29 Daar E, Kesler K, Wrin T, Petropoulos C, Bates M, Huang W, Hellmann N, Gomperts E, Donfield S and the Hemophila Growth and Development Study (2003) HIV co-receptor tropism (CRT) and replication capacity (RC) predict HIV progression. The 43rd Interscience Conference on Antimicrobial Agents and Chemotherapeutics, September 14–17, 2003, Chicago, IL, abstr. no. H-1722c

30 Van Rij RP, Blaak H, Visser JA et al (2000) Differential coreceptor expression allows for independent evolution of non-syncytium-inducing and syncytium-inducing HIV-1. *J Clin Invest* 106: 1569

31 Shepherd J, Jacobson L, Qiao W et al. (2006) Analysis of the timing of CXCR4-tropic HIV emergence in the MACS cohort. AIDS 2006 – XVI International AIDS Conference, Toronto, Canada, August 13–18, 2006, abstr. no. TUPE0001

32 Samson M, Libert F, Doranz BJ (1996) Resistance to HIV-1 infection in Caucasian individuals bearing mutant alleles of the CCR-5 chemokine receptor gene. *Nature* 382: 722–725

33 Glass WG, McDermott DH, Lim JK (2006) CCR5 deficiency increases risk of symptomatic West Nile virus infection. *J Exp Med* 203: 35–40

34 The AMD3100 HIV Study Group (2004) Safety, pharmacokinetics, and antiviral activity of AMD3100, a selective CXCR4 receptor inhibitor, in HIV-1 infection. *J Acquir Immune Defic Syndr* 37: 1253–1262

35 Mayer E, van der Ryst E, Saag M, Clotet B, Fätkenheuer G, Clumeck N, Turner K, Goodrich JM (2006) Safety and efficacy of Maraviroc (MVC), a novel CCR5 antagonist, when used in combination with optimized background therapy (OBT) for the treatment of antiretroviral-experienced subjects infected with dual/mixed-tropic HIV-1: 24-week results of a phase 2b exploratory trial. Program and abstracts of the XVI International AIDS Conference; August 13–18, 2006; Toronto, Ontario, abstr. no. THLB0215

36 Westby M, Whitcomb J, Huang W, James I, Abel S, Petropoulos C, Perros M, Van der Ryst E (2004) Reversible predominance of CXCR4 utilizing variants in a non-responsive dual tropic patient receiving the CCR5 antagonist UK-427,857. Abstracts of 11th Conference on Retroviruses and opportunistic Infections; February 8–11, 2004, San Francisco, CA, 538

37 Kitrinos K, LaBranche C, Stanhope M (2005) Clonal analysis detects pre-existing R5X4-tropic virus in patient demonstrating population-level tropism shift on 873140 monotherapy. *Antiviral Ther*10: S68

Future clinical prospects for entry inhibitors

Sonya L. Heath and J. Michael Kilby

Department of Medicine, Division of Infectious Diseases, 1917 Clinic, University of Alabama at Birmingham, 208 20th Street S., Birmingham, AL 35294, USA

Introduction

Some of the earliest attempts to develop HIV therapies involved agents intended to block viral entry into host cells, but only very recently and only once (with the FDA approval of a membrane fusion inhibitor, enfuvirtide, in 2003) has this strategic approach resulted in a commercially available agent. Indeed, for more than 15 years, from 1987 to 2003, all available antiretroviral therapies targeted one of two HW-encoded enzymes, reverse transcriptase (RT) or protease, which are critical components of later steps in the viral life cycle. Increasingly convenient combinations of RT and protease inhibitors (PI) have proven capable of potently suppressing viral replication and have dramatically improved the outlook for many HIV-infected patients. However, the sustained success of these enzyme inhibitors has been limited somewhat by selection for drug-resistant viral isolates, the necessity of strict dosing adherence, and the potential for toxicity. Thus, there remains a critical need for development of new therapeutic classes involving mechanisms of action distinctly different from RT and PI drugs. There is preliminary evidence demonstrating that viral entry inhibitors have potential to be safe and effective additions to the HIV armamentarium, and this class would be expected to have a low risk of cross-resistance with conventional antiretroviral drugs.

In this chapter, we compare and contrast the diverse entry inhibitors under development, or recently approved, in regard to clinically relevant characteristics (potency, pharmacokinetics, toxicity, resistance), with the clinical experience of currently available RT- and PI-based regimens. We describe some of the key lessons learned from the early clinical development of several HIV entry inhibitors, and discuss what these findings may mean in terms of future prospects, both promises and pitfalls, for this novel therapeutic class.

Potency

For preclinical and Phase I drug development, it is essential to demonstrate "proof of concept" that a novel approach shows potential for significant antiviral effects. Early in the HIV epidemic, there was a series of high profile investigative reports that generated considerable enthusiasm regarding the *in vitro* antiviral effects of soluble CD4, proposed to competitively inhibit binding of HIV particles to the CD4 receptors of target T helper cells [1, 2]. However, attempts to translate these findings into clinical protocols were generally disappointing, due to pharmacological constraints and an *in vitro* viral culture model that did not adequately reflect the complex *in vivo* situation [3, 4]. Indeed, these studies represent some of the earliest demonstrations that laboratory-adapted HIV isolates are not reliably predictive of HIV behavior in the clinical setting, whether in the context of developing an HIV vaccine or a new therapeutic strategy. Other nonspecific attachment inhibitors, such as dextran, also demonstrated *in vitro* antiviral potency [5] but proved potentially toxic and unpredictable in clinical pilot studies [6]. Inhibitors of the chemokine receptors, CCR5 and CXCR4, were identified in the laboratory even before the HIV/chemokine receptor nexus was well understood [7]. Unlike soluble CD4 and polyanionic compounds, CCR5 chemokine antagonists have shown clear evidence of potency in clinical pilot studies [8], but questions still remain about how well *in vitro* experience will reflect long-term efficacy against the complex viral quasispecies infecting a human host. A trial of AMD3100, a promising CXCR4 inhibitor *in vitro* [9], had no demonstrable antiviral activity except in one trial participant infected with a predominantly X4-tropic viral strain [10]. Since drug efficacy is confined by cellular tropism, there are concerns that a chemokine pathway inhibitor might predominantly provide selection pressures driving the evolution of viral tropism. For the CCR5 chemokine inhibitors, this would result in a relative shift from R5-tropic to (potentially more pathogenic) X4-tropic isolates, limiting the opportunity for sustained potent antiviral effects. Published papers demonstrate that these tropism shifts do occur, at least anecdotally, and more likely occur due to outgrowth of pre-existing viral populations rather than through individual viral isolates undergoing “tropism switching” [11]. Indeed, the primary modes of viral escape from CCR5 inhibition *in vitro* appear to involve mechanisms other than switching chemokine receptor tropism [12].

A recently reported study involving maraviroc, a CCR5 inhibitor, in treatment-experienced patients infected with mixed- or dual-tropic viruses, showed that in this clinical context there was little or no net effect on plasma viral load but there was a relative shift in the R5 and X4 tropic viral populations [13]. Interestingly, there was also a suggested immunological benefit, in that the CD4 count rose appreciably over the course of the study despite the lack of overall viral load suppression. Again there are signs that *in vitro* viral entry inhibition of selected viral strains does not necessarily extrapolate predictably into clinical benefits (but also the suggestion that benefits are possible in the absence of net plasma viral load reduction). It is possible that a combined assault on CCR5 and CXCR4, if this proves to be a safe clinical strategy, will be necessary to suppress viral replication more durably and consistently in the “real world” situation [14].

In the case of fusion inhibition, select regions of the gp41 sequence were found to be critically important for membrane fusion in the earliest years of

HIV scientific investigations [15]. Over subsequent years of study, peptides such as T20 (now formulated as enfuvirtide) were evaluated as possible therapeutics to interfere with fusion and entry. Perhaps because of the limited clinical successes with other viral entry inhibitors and the perceived narrow window of opportunity (peptides should only be capable of inhibiting the process during the moment of gp41 unfolding), there was understandable skepticism about how well this approach would translate into clinical studies. However, the initial short-term trial involving intravenous enfuvirtide monotherapy demonstrated dose-related viral suppression at a level of potency comparable to highly active antiretroviral regimens [16]. Subsequent trials involving subcutaneous administration provided encouragement about more sustained and practically achievable potency when combined with conventional agents for heavily treatment-experienced patients [17, 18]. Finally, two large Phase III protocols demonstrated that a salvage regimen including enfuvirtide plus the best combination of available agents ("optimized background") was significantly more potent than relying on optimized background alone [19, 20]. In part because of the nature of evaluating the drug as a "deep salvage" option, clinical data directly comparing the potency of enfuvirtide with other single agents such as efavirenz or a boosted PI are not available. Regardless of this, because enfuvirtide is relatively expensive and not available in an oral formulation, it will not be very acceptable as a first-line therapy (except perhaps in rather atypical clinical situations [21]). For pharmaceutical companies with entry inhibitor agents under development, this debate also leads to a critical question, whether to pursue indications based on the demonstration of antiviral potency in the setting of treatment-naïve *versus* treatment-experienced patients.

Two recent trials designed to compare an investigational oral CCR5 inhibitor with efavirenz as a component of first-line therapy are illustrative of this point. In 2005, a protocol comparing the CCR5 antagonist vicriviroc to efavirenz (each combined with two nucleoside RT inhibitors) in treatment-naïve patients was discontinued prematurely after an interim analysis suggested inferiority of vicriviroc to the standard of care first-line therapy [22]. More recently, the data safety monitoring board of another study involving the CCR5 antagonist maraviroc recommended discontinuation of the lower dose and more-convenient maraviroc arm (300 mg once daily) because preliminary analysis suggested that this arm was likely to be inferior to those in the efavirenz arm (Pfizer press release, 1/24/06). Beyond the "proof of concept" stage, defining the "potency" of viral entry inhibitors will obviously be impacted by the clinical context. It is quite possible, for example, for a drug to demonstrate clinically relevant advantages in the heavily pretreated subject with few other therapeutic options, while proving inferior to available treatments as a component of a first treatment regimen. This is a particularly relevant dilemma in the case of evaluating CCR5 antagonists, which theoretically might be advantageously positioned earlier in the sequence of HIV treatment regimens (since R5-tropic viruses are more typically transmitted from person to person, whereas X4-tropic viruses tend to evolve later in the course of disease) but which nonetheless will be in high demand for later-stage patients desiring access to novel treatments. It seems likely that a putative entry inhibitor must demonstrate substantial antiviral potency in early clinical trials (decreasing plasma viral load by, say, 100–1000-fold in a dose-dependent fashion) to warrant further development, but reaching the threshold to justify Phase III randomized trials may depend on more nuanced interpretations of what constitutes clinically relevant antiviral responses based on clinical context (e.g., prior treatment experience, drug-resistance patterns, ease and schedule of administration).

Pharmacokinetics and convenience

In the 1990s, there was more emphasis on relative antiviral potency as the key determinant of drug selection (which is more potent, zidovudine or didanosine? Nevirapine or indinavir?). Over the past 10 years or so, with significant increases in approved antiretroviral agents from the RT and PI classes, there has been a paradigm shift in what clinicians perceive as the high priority features that distinguish antiretroviral choices. When high-profile studies showed, for example, that lopinavir-ritonavir resulted in more favorable outcomes than nelfinavir [23] or that efavirenz was statistically superior to indinavir [24], the reason may have been that one agent was simply "more potent" than the other, but more likely convenience and tolerability played a major role in the achievement of better sustained results. To treating physicians, the reason for therapeutic superiority is almost a moot point – what matters most to them is not the pathway that led to these results but the conclusion that when all factors were combined, one drug (or one particular drug combination) resulted in more favorable outcomes than another. On the other hand, considerations regarding the "pathway to superiority" are sometimes important when the criticism is raised that the pool of participants qualifying and providing consent for a trial may not be a fair representation of what the "real world" experience will be like. For example, clinicians must decide whether the enrollment criteria and research environment for the Phase III TORO trials of enfuvirtide, an agent that involves significant drug administration costs and challenges, reasonably reflect the highly treatment-experienced patients in their own clinics. As with any clinical protocol, clinicians anticipate that results in the routine clinic setting may fall short of what was achieved in a carefully-controlled, highly motivated research subject population, but this discrepancy may be even more pronounced for a drug that must be administered by self-injection and therefore requires selection of relatively sophisticated, responsible, and hygienic patients. In regard to agents targeting the chemokine receptor interactions, clinicians will need to pay careful attention to the inclusion criteria for study protocols, particularly how receptor tropism was determined and defined, and take into account whether this type of screening will be applicable and feasible in routine settings.

Some of the earliest attempts to block HIV entry were thwarted, at least in part, by pharmacokinetic challenges. *In vitro* studies demonstrated significant potential for soluble CD4 as an inhibitor of intercellular spread of lab-adapted viruses, but it proved difficult to demonstrate evidence of *in vivo* antiviral activity except in the highest parenteral dose groups [3, 4]. Formulations of CD4 bound to antibodies or protein side-groups to prolong the half-life have also been explored, but still with minimal success [25]. Heparin-like polyanions had significant antiviral effects *in vitro* [5, 26], probably based on nonspecific charge interference with HIV binding to target cells, but this activity proved unachievable pharmacologically, even at intravenous doses associated with substantial hematological toxicity [6]. Preliminary data suggests that PRO-140, an anti-CCR5 monoclonal antibody (mAb), will block receptors and maintain an antiviral effect for a prolonged period after administration, suggesting the possibility of parenteral injections at intervals of many weeks [27]. Similarly, a pilot clinical trial of the anti-CD4 mAb, TNX-355, demonstrated that antiviral effects and CD4 count increases persisted weeks after a single dose [28].

An early investigational CXCR4 inhibitor, the bicyclam compound AMD-3100, was found to have potent *in vitro* antiviral activity against X4 viruses but was not orally bioavailable [29]. Fortunately, some other chemokine pathway inhibitors have shown recent promise as orally bioavailable small molecules, and this may turn out to be a pivotal breakthrough for the entry class as a whole. Both SCH 351125 (Schering C) and another CCR5 inhibitor, aplaviroc, although orally bioavailability [30, 31], were not further pursued due to potential toxicity. Vicriviroc (Schering D), another CCR5 antagonist in clinical trials [33], is dosed orally and is being pursued as a "stand-alone" formulation when given as part of a first antiretroviral regimen, and in the setting of treatment-experienced patients when given with ritonavir as a pharmacokinetic "booster". Although clinicians and patients have become accustomed to co-administering mini-dose ritonavir with selected protease inhibitors, especially for patients with extensive prior treatment experience, this characteristic would be a potential disadvantage for vicriviroc due to added cost, inconvenience, and drug-drug interaction considerations, compared with entry agents sufficiently bioavailable without boosting. The importance of this pharmacokinetic limitation is somewhat dependent on the larger question of whether and how we will ultimately utilize CCR5 inhibitors in the more advanced, heavily pretreated stages of HIV infection. Perhaps the most promising of these CCR5 small molecules currently is maraviroc (UK-427,857), which has demonstrated *in vitro* and short-term clinical activity; a formulation, which is dosed orally twice daily without boosting, is currently in clinical trials [8, 32].

The only viral entry agent currently approved by the FDA, enfuvirtide, is also the first antiretroviral agent available only in a parenteral form. A feature obviously unique to the peptide fusion inhibitor is its relative size (36 amino acids), much larger than any other antiretroviral chemical structures and indeed one of the most complex pharmaceutical compounds ever manufactured in mass quantities (covered in detail in Chapter 11). Like insulin, it is not feasible to administer this drug orally, as it would be quickly rendered inactive by oral and gastric enzymes. Since entry inhibitors do not require entry into the host cell, one potential advantage they have is fewer major drug-drug interactions. This is a problematic limitation of "boosted PI" approaches, currently emphasized in treatment-experience patients, because the mini-dose ritonavir "boost" used to potentiate other PI levels has complex combined inhibitory and inducing effects on the cytochrome p450 enzyme system that metabolizes many of the pharmaceutical treatments commonly used in General Medicine and Infectious Disease clinics. Although data are limited, membrane fusion inhibitor peptides also do not appear to require dose-adjustment for mild renal impairment.

While enfuvirtide remains a viable option for heavily pre-treated patients with limited oral treatment options, the marketplace will obviously favor compounds that are more practical for prolonged usage. Concerns about interpatient variability in drug levels and potential for drug resistance have limited the enthusiasm for pursuing once daily dosing of enfuvirtide. As discussed in Chapter 11, the possibility of a needle-less delivery system for enfuvirtide has the potential to improve feasibility and tolerability of parenteral administration. Another related peptide, T-1249, appeared promising in early phase trials, but is not being actively developed currently [33]. Other candidates for the next generation of peptide fusion inhibitors are also in pre-clinical development. Recently, two peptides derived from a gp41 HR2 sequence have shown promising *in vitro* potency (including activity against enfuvirtide- and T-1249-resistant strains) and favorable pharmacokinetics in non-human primate models [34]. These experimental peptides were synthesized with linkers and stabilizers to prolong serum half-life to the degree that once weekly subcutaneous dosing may be possible. Meanwhile, there are also orally bioavailable viral entry inhibitors, particularly the CCR5 inhibitors vicriviroc and maraviroc, that appear to be at least comparable (and possibly advantageous) to many currently available RT and PI agents in terms of pharmacokinetics and convenience.

Toxicity and tolerability

Although treatment-associated adverse events are incompletely understood, concerns have been raised about the possibility that viral enzyme inhibitors cause collateral damage to processes involving distantly related host enzymes. It has been proposed, for example, that RT inhibitors might impact host mitochondrial DNA transcription (contributing to neuropathy, myopathy, and hepatic toxicity) and PIs might alter functioning of human proteases (such as enzymes involved in glucose and lipid metabolism) [35–39]. A theoretical safety advantage of viral entry inhibitors therefore is that these diverse agents are not enzyme inhibitors at all. However, it is important to keep in mind the lessons repeatedly learned from the RT and PI classes: (1) adverse events may not come to full clinical attention until after the drugs have been on the market for many years, and (2) the exact incidence and mechanisms of various adverse events may be hard to parse out from the effects of HIV itself as well as the effects of co-administered antiretroviral agents. Thus, any generalizations about the short-term toxicities of viral entry inhibitors must be interpreted carefully with these points in mind.

There are concerns regarding agents that target host receptor interactions integral to normal inflammatory and immune responses. Although soluble CD4 approaches lacked potency, they were well tolerated without substantial toxicities. However, some efforts to target CD4 with mAb in other areas of medicine, for example compounds to ameliorate immune-mediated rheumatological disorders, have shown potential to result in prolonged depletion of $CD4^+$ lymphocytes [40]. At the other immunological extreme, a recent study involving normal volunteers taking a single test dose of an anti-CD28 mAb, designed to modulate T cell immunoactivation, resulted in dramatic and unexpected consequences, essentially a life-threatening cytokine storm and multi-organ failure [41]. These experiences, while strictly speaking outside the HIV research field, serve as reminders about potential unanticipated consequences of manipulating the complex milieu of inflammatory and immune trafficking pathways.

Because the role of chemokine receptors in natural inflammatory and immune responses is not fully understood, the impact of manipulating these systems is considered uncharted territory. There is an experiment of nature, in effect, for the CCR5 receptor, in that individuals who lack gene expression of this receptor appear to have a full lifespan and no obvious immunological defects. The subset of individuals who are homozygous recessive for CCR5 receptor expression are over-represented among the occasionally identified "highly exposed but non-infected" individuals, relative to the general population, an observation that first led to the recognition of the key role of CCR5 for HIV entry [42]. Recent observations suggest the possibility that host deficiency of CCR5 expression, a protective state in terms of HIV pathogenesis, may increase the risk of encephalitis caused by West Nile virus [43], and alter the risks for immune-mediated diseases such as multiple sclerosis [44]. Again, it would appear prudent to monitor closely for unanticipated, subtle and delayed adverse outcomes in these chemokine antagonist studies. There is less subtlety regarding the importance of the other chemokine receptor commonly utilized by some HIV strains, CXCR4, for normal development and survival. Mice with deletions of CXCR4 die *in utero*, probably due to defects in cellular migration signals necessary for normal tissue and organ maturation [45]. Thus far, however, there have not been substantial or widespread toxicities relating to immune function or inflammation noted in the early chemokine antagonist studies. An early CCR5 inhibitor, SCH C, showed favorable activity but was associated with unacceptable cardiac effects including dose-dependent QT prolongation [30]. Another CCR5 inhibitor, SCH D (vicriviroc), lacks the chemical structure associated with the arrhythmia risk, and QT prolongation has not been demonstrated to date [46]. There has been some discussion about five cases of incident malignancies (lymphomas, gastric carcinoma) occurring among vicriviroc recipients but it remains unclear whether these findings have any relationship with the study drug (see Chapter 12). Aplaviroc development was halted in late 2005 due to severe hepatotoxicity in at least 5 subjects [47], but thus far this has not been observed with vicroviroc or maraviroc, suggesting that this may not be a broad class effect. Nevertheless, longer-term follow-up in larger clinical trials will be important.

The other broad category of theoretical toxicity associated with chemokine inhibition relates to undesirable effects on the infecting viral population. In addition to limiting the potency of the agent for subjects infected with mixed or dual-tropic viruses, potent CCR5 inhibition also raises the possibility of providing selection pressures that drive the viral quasispecies towards a more pathogenic phenotype. There are insufficient data to determine whether this kind of selective pressure has any reasonable chance of adversely affecting the natural history of disease, but this potential will continue to influence the design of clinical trials and the strategic planning for long-term monitoring. Combined inhibition of CCR5 and CXCR4 receptors, while diminishing chances for a skewed antiviral selection process, might also raise the potential for toxicities relating to host immune and inflammatory pathways. One potential solution is to develop compounds that selectively block HIV engagement at chemokine receptor sites while allowing natural ligands to bind without interference. Because the anti-CD4 mAb, TNX-355, is proposed to work by sterically inhibiting the binding of HIV to adjacent chemokine receptors, this might represent a method of blocking a chemokine pathway with less impact on other aspects of the immune response [28].

Systemic reactions or significant end-organ damage have been relatively uncommon with enfuvirtide and related peptide membrane fusion inhibitors thus far. In contrast to many of the RT and PI agents, trials have not suggested a strong association with gastrointestinal upset, diarrhea, hyperlipidemia, or neuropathy (at least substantial enough to be noticeable above rates seen with the "optimized background" regimens in the TORO studies). Instead, most concerns about toxicity and tolerability have been related to the route of administration, particularly the nearly universal "injection site reactions" (ISR) and the potential for patient "burnout" after rotating injection sites twice daily over long periods of time. While most ISR have been mild to moderate and not resulted in drug discontinuation, there have been rare systemic anaphylactic-type reactions, proven to be authentic by carefully monitored rechallenges. Biopsy studies and long-term follow-up of ISR suggest that most are relatively minor, with nonspecific pathological findings [48, 49], and do not have lasting consequences for most subjects.

One intriguing new research direction is the evaluation of a needle-less air-injection device, the Biojector B2000 (Bioject Inc., Portland, OR), which may improve the safety and tolerability of enfuvirtide administration. This portable gas-powered system forces medication rapidly through the skin to disperse it into subcutaneous tissues, potentially with less tissue trauma and therefore lesser propensity for bleeding and pain than a needle. An observational study, involving 32 HIV-infected volunteers who had already experienced ISR to enfuvirtide, assessed subjects before and after switching from standard needle injections to the Biojector gas-powered injection system [50]. Drug levels were not statistically different during the two phases (gas-powered or needle-injected) of the study, either at the pre-dose trough or the 1-h post-dose time points. Based on a predetermined scale related to ease of use, participants rated the Biojector approach significantly better than standard administration. A separate injection site rating instrument (which was also used in the TORO trials), combining a subjective self-reported score and a graded scale for objective findings also demonstrated a significant advantage in ISR scores for the Biojector *versus* standard needle administration. The majority of subjects opted to continue with the needle-free system after the initial comparison phase was completed. A larger, multicenter trial to further evaluate gas-powered subcutaneous administration of enfuvirtide is underway in 2007.

In summary, while several unanswered questions persist in regard to the long-term safety of viral entry inhibitors, the class has come through thus far without evidence for substantial "overlapping toxicities" with the RT and PI classes. Whereas chemokine and fusion inhibitors introduce new potential toxicity concerns, these agents remain viable options for patients who have limited treatment options due to adverse events or gastrointestinal intolerability associated with conventional antiretroviral drugs, and these or future entry inhibitors may well prove to be equal or advantageous compared with RT and PI drugs for first-line therapy as well (see Tab. 1).

Resistance

Viral entry inhibitors in general are unique in that the site of action is not dependent on cell entry. Whereas there are proposed mechanisms for RT and PI resistance at the host level, due to modulation of cellular efflux transporter function [51, 52], these considerations are generally not relevant to agents with entirely extracellular mechanisms of action. Investigational agents targeting highly variable portions of the viral envelope may prove to have daunting limitations due to the relative frequency and seemingly endless variety of possible escape mechanisms. BMS-806, a small molecule targeting the gp120-CD4 interaction, demonstrated potent antiviral effects *in vitro*, but activity may be quite variable between different viral isolates, and drug sensitivity may be affected in a complex fashion by changes in envelope configuration [53]. Resistance concerns, however, are not limited to drugs with highly variable viral targets. TNX-355, the anti-CD4 mAb, was mentioned previously in regard to the prolonged antiviral effect following a single dose; however, recent data suggest that the majority of subjects had significant reductions in drug activity over the short course of the study [54].

Escape from chemokine antagonists can occur via viral mutations resulting in altered tropisms or cellular changes. Earlier experiments involving

Table 1. Selected characteristics of some approved and investigational antiretroviral agents, to broadly compare reverse transcriptase and protease inhibitors with viral entry inhibitors

Class	PK	Examples of toxicities	Resistance
NRTI	– Oral, once or twice daily – Combination formulations – Limited drug interactions – Some pairings in regimen "backbones" not compatible (DDI/D4T; ABC/Tenofovir; etc)	– Neuropathy (DDI; D4T) – Hypersensitivity (ABC) – Lipoatrophy, hyperlipidemia (d4T) – Anemia (AZT) – Pancreatitis (DDI) **– Hepatotoxicity, hyperlactatemia (many or all?)**	– Complex, stepwise RCMs – Partial cross-class considerations (TAMS)
NNRTI	– Oral, once or twice daily – Some drug interactions – Limited renal clearance concerns	– Rash, potentially severe, hepatotoxicity (NVP) – CNS effects (EFV)	**– Low genetic barrier** **– Nearly complete cross-class resistance**, but new generation available soon
PI	– Oral, once or twice daily **– Boosting recommended** **– Diverse absorption and drug interaction issues**	**– Diarrhea, nausea, hyperlipidemia, insulin resistance (many or all?)**	– High genetic barrier – Complex, stepwise RCMs – Increasing opportunities for sequencing
Chemokines	– Oral, once or twice daily – Limited drug interactions? – Boosting sometimes necessary?	– Arrythmia (SCH C) – Hepatotoxicity (aplaviroc) – Theoretical concerns about immune responses/inflammatory pathways (many or all?)	**– Concerns about tropism selection pressures** – Unclear role of CCR5 inhibitors in late-stage disease? – Escape mutations?
Fusion Inhibitors	**– Subcutaneous self-injections**, twice daily – Drug interactions uncommon	**– Infusion site reactions**	**– Low genetic barrier**

Some key comparative features described in this chapter are in bold type.
DDI: didanosine; D4T: stavudine; ABC: abacavir; AZT: zidovudine; NVP: nevirapine; EFV: efavirenz; RCMs: resistance-conferring mutations; TAMS: thymidine analog resistance mutations.

chemokine ligands as *in vitro* antiretroviral compounds, presumably acting as competitive inhibitors against HIV binding, demonstrated a resulting down-regulation of chemokine coreceptor expression as at least a partial explanation for their activity [55]. Evidence suggests that maraviroc may not result in altered expression of CCR5 on the cell surface as a potential escape mechanism, which may also be reassuring in terms of potential toxicity consequences [32]. However, there are anecdotal examples of apparent escape by virtue of viral "population dynamics" in response to drug pressures, and this will remain a phenomenon that warrants close attention in future clinical trials. Although there are limited data about the clinical significance of drug-resistant isolates in relation to chemokine receptor antagonists, lab-adapted strains have demonstrated cross resistance to several types of CCR5 inhibitors while retaining the R5 phenotype [12]. Such resistance is not limited to genetic mutations within the chemokine receptor region, but may also be impacted by complex alterations throughout the variable loops (V2, V3, V4) of the viral envelope [56]. Maraviroc has *in vitro* activity against a wide variety of HIV isolates and lab-adapted strains, regardless of resistance to other antiviral classes such as RT and PI, and there is generally not a wide range of susceptibility (~10-fold or less) from isolate to isolate as has been reported with the membrane fusion inhibitors [32].

As with most or all of the RT and PI antiretroviral agents, the membrane fusion inhibitors are associated with characteristic resistance-conferring mutations that have been demonstrated *in vitro* and *in vivo*. Early studies involving *in vitro* passage of HIV-1 isolates in the presence of increasing enfuvirtide concentrations demonstrated selection for mutations in the first heptad repeat (HR-1) region of gp41 where the peptide was proposed to bind, and these observations provided confirmation of the putative membrane fusion mechanism of action [57]. These and later *in vitro* experiments consistently noted mutations occurring in the gp41 sequence of amino acids between 36 and 45, and in particular the relatively conserved 36–38 ("GIV") sequence [58–60]. Various mutations in the GIV region (36S, 36D, 37 T, 38 M) contribute to enfuvirtide resistance (often in the range of 10-fold), and the combined presence of two mutations (at positions 36 and 38 for example) may increase phenotypic resistance >100-fold.

Evidence suggests there is a low genetic barrier for rapid selection of enfuvirtide resistance, particularly when it is administered at suboptimal doses or without adequate additional potent agents. Typically individuals with clinical drug failure have viral isolates with identifiable mutations in the HR-1 sequence. In the TORO protocols, 93% of subjects who met protocol criteria for virological failure had demonstrable changes from pre-therapy baseline within the 36–45 gp41 sequence, and this occurred in virtually all subjects who had a 4-fold change in phenotypic enfuvitide resistance [61]. Although there was a wide variation in pre-treatment phenotypic enfuvirtide susceptibilities (despite wild-type GIV sequences in all cases at baseline), as reported previously, having a lower pre-therapy susceptibility did not predict a poor response to therapy. Other studies have shown that if the fusion inhibitor is continued long-term in this setting of rising viral load during therapy there is evidence of ongoing selection for other resistance-conferring mutations in the HR-2 region and elsewhere, possibly compensatory changes that no longer compromise fitness to the same degree [62, 63]. As discussed above, preliminary data exist to support sequencing of fusion inhibitors, in that newer generation peptides have activity against enfuvirtide-resistant viruses [33, 34].

Thus, as with all antiretroviral agents studied to date, the selection of viral entry inhibitor-resistant strains appears to be frustratingly rapid, common and quite complex in the right scenarios. This diverse class of novel entry inhibitors has not dramatically upstaged other available agents in terms of drug resistance based on preliminary evidence, but as with toxicity and tolerability considerations at least there is little to suggest substantial overlap with resistance determinants associated with the RT and PI classes. Clearly, the same principles of careful adherence, as well as rational drug selection and regimen sequencing, will need to be maintained with the viral entry inhibitors as with the pre-existing therapies if we are to derive optimal benefits from each of these promising agents.

Conclusions

Perhaps the most favorable aspect to consider in regard to the viral entry inhibitors is the diversity of potential targets and step-wise pathways that appear to warrant further exploration. Compared with the conventional RT and PI enzyme inhibitors, the growing list of entry inhibition strategies may actually represent multiple novel "treatment classes". Laboratory investigations have suggested unprecedented degrees of synergism when entry inhibitors with different mechanisms of action are combined *in vitro*. Currently, early trials are underway which for the first time will evaluate the clinical efficacy of combining chemokine inhibitors and enfuvirtide. In fact, it is quite conceivable that the relatively near future will hold the prospect of an "all entry inhibitor" highly active regimen that, when sequenced (whether first, last, or in the middle) alongside RT- and PI-based regimens, will expand treatment options substantially and thereby stretch the durability of overall antiretroviral treatment effectiveness and the lifespan of HIV-infected patients.

With the diversity of entry inhibition strategies in mind, it can be somewhat misleading to lump these compounds together and rate how they compare with current treatment options. Primarily it is the chemokine receptor antagonists and membrane fusion inhibitors that have generated sufficient data to formulate some broad comparisons to the traditional HIV enzyme inhibitors. *In vitro* antiviral potency has been demonstrated for a variety of different entry inhibitors, but consistent long-term clinical benefits remain in question, except in the case of using enfuvirtide to supplement best available regimens for heavily pre-treated patients. Although it is beyond the scope of the current chapter, it is important to remember that several entry inhibitor strategies also have unique potential applications for the prevention of HIV infection, for example in the form of post-exposure prophylaxis or vaginal microbicides. The major liabilities for the currently available systemic viral entry agent, enfuvirtide, are the costs and inconvenience of parenteral administration, but there are signs of improvements and alternatives on the horizon, in the form of longer-acting or more user friendly delivery of fusion inhibitors as well as the promise of orally bioavailable small molecule entry inhibitors with other mechanisms of action. Although clinical experience remains relatively limited, enfuvirtide and several of the other viral entry inhibitor agents in development appear to share advantages over the PI class in terms of adverse drug-drug interactions and overall tolerability. However, the almost universal occurrence of injection site reactions with enfuvirtide, the arrhythmic effects of SCH-C, and the hepatotoxicity that halted trials involving aplaviroc serve as reminders for due diligence as we monitor for unexpected developments in the future. The selection for drug resistance is likely to be equally problematic for these agents as for RT and PI drugs, and manipulating the chemokine nexus adds a new layer of concerns regarding immunological safety as well as the consequences of inducing population shifts in viral tropism. A common theme for several of these categorical comparisons is that, whereas none of these new agents have achieved optimal performance characteristics, there has been significant success in avoiding additive or overlapping toxicities, intolerable effects, pharmacokinetic challenges, and resistance problems with currently existing therapies. This is not a minor achievement, as the addition of these novel agents will expand our abilities to individualize therapeutic options, improving our ability to adapt to the needs of the diverse individuals affected by the expanding HIV epidemic.

References

1 Fisher RA, Bertonis JM, Meier W, Johnson VA, Costopoulos DS, Liu T, Tizard R, Walker BD, Hirsch MS, Schooley RT et al. (1988) HIV infection is blocked *in vitro* by recombinant soluble CD4. *Nature* 331: 76–78

2 Hussey RE, Richardson NE, Kowalski M, Brown NR, Chang HC, Siliciano RF, Dorfman T, Walker B, Sodroski J, Reinherz EL (1988) A soluble CD4 protein selectively inhibits HIV replication and syncytium formation. *Nature* 331: 78–81

3 Schooley RT, Merigan TC, Gaut P, Hirsch MS, Holodniy M, Flynn T, Liu S, Byington RE, Henochowicz S, Gubish E et al. (1990) Recombinant soluble CD4 therapy in patients with the acquired immunodeficiency syndrome (AIDS) and AIDS-related complex. A phase I-II escalating dosage trial. *Ann Intern Med* 112: 247–253

4 Schacker T, Collier AC, Coombs R, Unadkat JD, Fox I, Alam J, Wang JP, Eggert E, Corey L (1995) Phase I study of high-dose, intravenous rsCD4 in subjects with advanced HIV-1 infection. *J Acquir Immune Defic Syndr Hum Retrovirol* 9: 145–152

5 Mitsuya H, Looney DJ, Kuno S, Ueno R, Wong-Staal F, Broder S (1988) Dextran sulfate suppression of viruses in the HIV family: inhibition of virion binding to $CD4^+$ cells. *Science* 240: 646–649

6 Flexner C, Barditch-Crovo PA, Kornhauser DM, Farzadegan H, Nerhood LJ, Chaisson RE, Bell KM, Lorentsen KJ, Hendrix CW, Petty BG et al. (1991) Pharmacokinetics, toxicity, and activity of intravenous dextran sulfate in human immunodeficiency virus infection. *Antimicrob Agents Chemother* 35: 2544–2550

7 Cocchi F, DeVico AL, Garzino-Demo A, Arya SK, Gallo RC, Lusso P (1995) Identification of RANTES, MIP-1 alpha, and MIP-1 beta as the major HIV-suppressive factors produced by CD8+ T cells. *Science* 270: 1811–1815

8 Fatkenheuer G, Pozniak AL, Johnson MA, Plettenberg A, Staszewski S, Hoepelman AI, Saag MS, Goebel FD, Rockstroh JK, Dezube BJ et al. (2005) Efficacy of short-term monotherapy with maraviroc, a new CCR5 antagonist, in patients infected with HIV-1. *Nat Med* 11: 1170–1172

9 Donzella GA, Schols D, Lin SW, Este JA, Nagashima KA, Maddon PJ, Allaway GP, Sakmar TP, Henson G, De Clercq E et al. (1998) AMD3100, a small molecule inhibitor of HIV-1 entry via the CXCR4 co-receptor. *Nat Med* 4: 72–77

10 Hendrix CW, Collier AC, Lederman MM, Schols D, Pollard RB, Brown S, Jackson JB, Coombs RW, Glesby MJ, Flexner CW et al. (2004) Safety, pharmacokinetics, and antiviral activity of AMD3100, a selective CXCR4 receptor inhibitor, in HIV-1 infection. *J Acquir Immune Defic Syndr* 37: 1253–1262

11 Westby M, Lewis M, Whitcomb J, Youle M, Pozniak AL, James IT, Jenkins TM, Perros M, van der Ryst E (2006) Emergence of CXCR4-using human immunodeficiency virus type 1 (HIV-1) variants in a minority of HIV-1-infected patients following treatment with the CCR5 antagonist maraviroc is from a pretreatment CXCR4-using virus reservoir. *J Virol* 80: 4909–4920

12 Trkola A, Kuhmann SE, Strizki JM, Maxwell E, Ketas T, Morgan T, Pugach P, Xu S, Wojcik L, Tagat J et al. (2002) HIV-1 escape from a small molecule, CCR5-specific entry inhibitor does not involve CXCR4 use. *Proc Natl Acad Sci USA* 99: 395–400

13 Mayer H, Rest EV, Saag MS (2006) Safety and efficacy of maraviroc, a novel CCR5 antagonist, when used in combination with optimised background therapy for the treatment of antiretroviral-experienced subjects infected with dual/mixed tropic HIV-1: 24-week results of a phase 2b exploratory trial. International AIDS Society Toronto, THLB-0215

14 Kilby JM, Eron JJ (2003) Novel therapies based on mechanisms of HIV-1 cell entry. *N Engl J Med* 348: 2228–2238

15 Dubay JW, Roberts SJ, Brody B, Hunter E (1992) Mutations in the leucine zipper of the human immunodeficiency virus type 1 transmembrane glycoprotein affect fusion and infectivity. *J Virol* 66: 4748–4756

16 Kilby JM, Hopkins S, Venetta TM, DiMassimo B, Cloud GA, Lee JY, Alldredge L, Hunter E, Lambert D, Bolognesi D et al. (1998) Potent suppression of HIV-1 replication in humans by T-20, a peptide inhibitor of gp41-mediated virus entry. *Nat Med* 4: 1302–1307

17 Kilby JM, Lalezari JP, Eron JJ, Carlson M, Cohen C, Arduino RC, Goodgame JC, Gallant JE, Volberding P, Murphy RL et al. (2002) The safety, plasma pharmacokinetics, and antiviral activity of subcutaneous enfuvirtide (T-20), a peptide inhibitor of gp41-mediated virus fusion, in HIV-infected adults. *AIDS Res Hum Retroviruses* 18: 685–693

18 Lalezari JP, Eron JJ, Carlson M, Cohen C, DeJesus E, Arduino RC, Gallant JE, Volberding P, Murphy RL, Valentine F et al. (2003) A phase II clinical study of the long-term safety and antiviral activity of enfuvirtide-based antiretroviral therapy. *AIDS* 17: 691–698

19 Lalezari JP, Henry K, O'Hearn M, Montaner JS, Piliero PJ, Trottier B, Walmsley S, Cohen C, Kuritzkes DR, Eron JJ Jr et al. (2003) Enfuvirtide, an HIV-1 fusion inhibitor, for drug-resistant HIV infection in North and South America. *N Engl J Med* 348: 2175–2185

20 Lazzarin A, Clotet B, Cooper D, Reynes J, Arasteh K, Nelson M, Katlama C, Stellbrink HJ, Delfraissy JF, Lange J et al. (2003) Efficacy of enfuvirtide in patients infected with drug-resistant HIV-1 in Europe and Australia. *N Engl J Med* 348: 2186–2195

21 Bourgarit A, Lascoux C, Palmer P, Tuleja E, Pintado C, Farge D, Sereni D (2006) First-line use of enfuvirtide-containing HAART regimen with dramatic clinical and immunological improvement in three cases. *AIDS* 20: 471–473

22 Greaves W, R L, Fatkenheuer G (2006) Late virologic breakthrough in treatment-naive patients on a regimen of Combivir + viriviroc. 13th Conference on Retroviruses and Opportunistic Infections, Denver, CO

23 Walmsley S, Bernstein B, King M, Arribas J, Beall G, Ruane P, Johnson M, Johnson D, Lalonde R, Japour A et al. (2002) Lopinavir-ritonavir *versus* nelfinavir for the initial treatment of HIV infection. *N Engl J Med* 346: 2039–2046

24 Staszewski S, Morales-Ramirez J, Tashima KT, Rachlis A, Skiest D, Stanford J, Stryker R, Johnson P, Labriola DF, Farina D et al. (1999) Efavirenz plus zidovudine and lamivudine, efavirenz plus indinavir, and indinavir plus zidovudine and lamivudine in the treatment of HIV-1 infection in adults. Study 006 Team. *N Engl J Med* 341: 1865–1873

25 Meng TC, Fischl MA, Cheeseman SH, Spector SA, Resnick L, Boota A, Petrakis T, Wright B, Richman DD (1995) Combination therapy with recombinant human soluble CD4-immunoglobulin G and zidovudine in patients with HIV infection: a phase I study. *J Acquir Immune Defic Syndr Hum Retrovirol* 8: 152–160

26 Baba M, Pauwels R, Balzarini J, Arnout J, Desmyter J, De Clercq E (1988) Mechanism of inhibitory effect of dextran sulfate and heparin on replication of human immunodeficiency virus *in vitro*. *Proc Natl Acad Sci USA* 85: 6132–6136

27 Castagna A, Biswas P, Beretta A, Lazzarin A (2005) The appealing story of HIV entry inhibitors: from discovery of biological mechanisms to drug development. *Drugs* 65: 879–904

28 Kuritzkes DR, Jacobson J, Powderly WG, Godofsky E, DeJesus E, Haas F, Reimann KA, Larson JL, Yarbough PO, Curt V et al. (2004) Antiretroviral activity of the anti-CD4 monoclonal antibody TNX-355 in patients infected with HIV type 1. *J Infect Dis* 189: 286–291

29 Hendrix CW, Flexner C, MacFarland RT, Giandomenico C, Fuchs EJ, Redpath E, Bridger G, Henson GW (2000) Pharmacokinetics and safety of AMD-3100, a novel antagonist of the CXCR-4 chemokine receptor, in human volunteers. *Antimicrob Agents Chemother* 44: 1667–1673

30 Strizki JM, Xu S, Wagner NE, Wojcik L, Liu J, Hou Y, Endres M, Palani A, Shapiro S, Clader JW et al. (2001) SCH-C (SCH 351125), an orally bioavailable, small molecule antagonist of the chemokine receptor CCR5, is a potent inhibitor of HIV-1 infection *in vitro* and *in vivo*. *Proc Natl Acad Sci USA* 98: 12718–12723

31 Adkison KK, Shachoy-Clark A, Fang L, Lou Y, O'Mara K, Berrey MM, Piscitelli SC (2005) Pharmacokinetics and short-term safety of 873140, a novel CCR5 antagonist, in healthy adult subjects. *Antimicrob Agents Chemother* 49: 2802–2806

32 Dorr P, Westby M, Dobbs S, Griffin P, Irvine B, Macartney M, Mori J, Rickett G, Smith-Burchnell C, Napier C et al. (2005) Maraviroc (UK-427,857), a potent, orally bioavailable, and selective small-molecule inhibitor of chemokine receptor CCR5 with broad-spectrum anti-human immunodeficiency virus type 1 activity. *Antimicrob Agents Chemother* 49: 4721–4732

33 Eron JJ, Gulick RM, Bartlett JA, Merigan T, Arduino R, Kilby JM, Yangco B, Diers A, Drobnes C, DeMasi R et al. (2004) Short-term safety and antiretroviral activity of T-1249, a second-generation fusion inhibitor of HIV. *J Infect Dis* 189: 1075–1083

34 Delmedico M, Bray B, Cammack N (2006) Next generation HIV peptide fusion inhibitor candidates achieve potent, durable suppression of virus replication *in vitro* and improved pharmacokinetic properties. 13th Conference on Retroviruses and Opportunistic Infections, Denver, CO

35 Cote HC, Yip B, Asselin JJ, Chan JW, Hogg RS, Harrigan PR, O'Shaughnessy MV, Montaner JS (2003) Mitochondrial:nuclear DNA ratios in peripheral blood cells from human immunodeficiency virus (HIV)-infected patients who received selected HIV antiretroviral drug regimens. *J Infect Dis* 187: 1972–1976

36 Montaner JS, Cote HC, Harris M, Hogg RS, Yip B, Chan JW, Harrigan PR, O'Shaughnessy MV (2003) Mitochondrial toxicity in the era of HAART: evaluating venous lactate and peripheral blood mitochondrial DNA in HIV-infected patients taking antiretroviral therapy. *J Acquir Immune Defic Syndr* 34 Suppl 1: S85–90

37 Carr A, Cooper DA (2000) Adverse effects of antiretroviral therapy. *Lancet* 356: 1423–1430

38 Nolan D (2003) Metabolic complications associated with HIV protease inhibitor therapy. *Drugs* 63: 2555–2574

39 Powderly WG (2002) Long-term exposure to lifelong therapies. *J Acquir Immune Defic Syndr* 29 Suppl 1: S28–40

40 Moreland LW, Bucy RP, Koopman WJ (1995) Regeneration of T cells after chemotherapy. *N Engl J Med* 332: 1651–1652

41 Suntharalingam G, Perry MR, Ward S, Brett SJ, Castello-Cortes A, Brunner MD, Panoskaltsis N (2006) Cytokine storm in a phase 1 trial of the anti-CD28 monoclonal antibody TGN1412. *N Engl J Med* 355: 1018–1028

42 Liu R, Paxton WA, Choe S, Ceradini D, Martin SR, Horuk R, MacDonald ME, Stuhlmann H, Koup RA, Landau NR (1996) Homozygous defect in HIV-1 coreceptor accounts for resistance of some multiply-exposed individuals to HIV-1 infection. *Cell* 86: 367–377

43 Glass WG, McDermott DH, Lim JK, Lekhong S, Yu SF, Frank WA, Pape J, Cheshier RC, Murphy PM (2006) CCR5 deficiency increases risk of symptomatic West Nile virus infection. *J Exp Med* 203: 35–40

44 Favorova OO, Andreewski TV, Boiko AN, Sudomoina MA, Alekseenkov AD, Kulakova OG, Slanova AV, Gusev EI (2002) The chemokine receptor CCR5 deletion mutation is associated with

MS in HLA-DR4-positive Russians. *Neurology* 59: 1652–1655
45 Nagasawa T, Hirota S, Tachibana K, Takakura N, Nishikawa S, Kitamura Y, Yoshida N, Kikutani H, Kishimoto T (1996) Defects of B-cell lymphopoiesis and bone-marrow myelopoiesis in mice lacking the CXC chemokine PBSF/SDF-1. *Nature* 382: 635–638
46 Strizki JM, Tremblay C, Xu S, Wojcik L, Wagner N, Gonsiorek W, Hipkin RW, Chou CC, Pugliese-Sivo C, Xiao Y et al. (2005) Discovery and characterization of vicriviroc (SCH 417690), a CCR5 antagonist with potent activity against human immunodeficiency virus type 1. *Antimicrob Agents Chemother* 49: 4911–4919
47 Deeks SG (2006) Challenges of developing R5 inhibitors in antiretroviral naive HIV-infected patients. *Lancet* 367: 711–713
48 Ball RA, Kinchelow T (2003) Injection site reactions with the HIV-1 fusion inhibitor enfuvirtide. *J Am Acad Dermatol* 49: 826–831
49 Maggi P, Ladisa N, Cinori E, Altobella A, Pastore G, Filotico R (2004) Cutaneous injection site reactions to long-term therapy with enfuvirtide. *J Antimicrob Chemother* 53: 678–681
50 Harris M, Joy R, Larsen G, Valyi M, Walker E, Frick LW, Palmatier RM, Wring SA, Montaner JS (2006) Enfuvirtide plasma levels and injection site reactions using a needle-free gas-powered injection system (Biojector). *AIDS* 20: 719–723
51 Fridland A, Connelly MC, Robbins BL (2000) Cellular factors for resistance against antiretroviral agents. *Antiviral Ther* 5: 181–185
52 Lucia MB, Rutella S, Leone G, Vella S, Cauda R (2001) HIV-protease inhibitors contribute to P-glycoprotein efflux function defect in peripheral blood lymphocytes from HIV-positive patients receiving HAART. *J Acquir Immune Defic Syndr* 27: 321–330
53 Madani N, Perdigoto AL, Srinivasan K, Cox JM, Chruma JJ, LaLonde J, Head M, Smith AB 3rd, Sodroski JG (2004) Localized changes in the gp120 envelope glycoprotein confer resistance to human immunodeficiency virus entry inhibitors BMS-806 and #155. *J Virol* 78: 3742–3752
54 Duensing T, Fung M, Lewis S (2006) *In vitro* characterization of HIV isolated from patients treated with the entry inhibitor TNX-355. 13th Conference on Retroviruses and Opportunistic Infections, Denver, CO
55 Amara A, Gall SL, Schwartz O, Salamero J, Montes M, Loetscher P, Baggiolini M, Virelizier JL, Arenzana-Seisdedos F (1997) HIV coreceptor downregulation as antiviral principle: SDF-1alpha-dependent internalization of the chemokine receptor CXCR4 contributes to inhibition of HIV replication. *J Exp Med* 186: 139–146
56 Briz V, Poveda E, Soriano V (2006) HIV entry inhibitors: mechanisms of action and resistance pathways. *J Antimicrob Chemother* 57: 619–627
57 Rimsky LT, Shugars DC, Matthews TJ (1998) Determinants of human immunodeficiency virus type 1 resistance to gp41-derived inhibitory peptides. *J Virol* 72: 986–993
58 Derdeyn CA, Decker JM, Sfakianos JN, Wu X, O'Brien WA, Ratner L, Kappes JC, Shaw GM, Hunter E (2000) Sensitivity of human immunodeficiency virus type 1 to the fusion inhibitor T-20 is modulated by coreceptor specificity defined by the V3 loop of gp120. *J Virol* 74: 8358–8367
59 Derdeyn CA, Decker JM, Sfakianos JN, Zhang Z, O'Brien WA, Ratner L, Shaw GM, Hunter E (2001) Sensitivity of human immunodeficiency virus type 1 to fusion inhibitors targeted to the gp41 first heptad repeat involves distinct regions of gp41 and is consistently modulated by gp120 interactions with the coreceptor. *J Virol* 75: 8605–8614
60 Mink M, Mosier SM, Janumpalli S, Davison D, Jin L, Melby T, Sista P, Erickson J, Lambert D, Stanfield-Oakley SA et al. (2005) Impact of human immunodeficiency virus type 1 gp41 amino acid substitutions selected during enfuvirtide treatment on gp41 binding and antiviral potency of enfuvirtide *in vitro*. *J Virol* 79: 12447–12454
61 Melby T, Sista P, DeMasi R, Kirkland T, Roberts N, Salgo M, Heilek-Snyder G, Cammack N, Matthews TJ, Greenberg ML (2006) Characterization of envelope glycoprotein gp41 genotype and phenotypic susceptibility to enfuvirtide at baseline and on treatment in the phase III clinical trials TORO-1 and TORO-2. *AIDS Res Hum Retroviruses* 22: 375–385
62 Xu L, Pozniak A, Wildfire A, Stanfield-Oakley SA, Mosier SM, Ratcliffe D, Workman J, Joall A, Myers R, Smit E et al. (2005) Emergence and evolution of enfuvirtide resistance following long-term therapy involves heptad repeat 2 mutations within gp41. *Antimicrob Agents Chemother* 49: 1113–1119
63 Menzo S, Castagna A, Monachetti A, Hasson H, Danise A, Carini E, Bagnarelli P, Lazzarin A, Clementi M (2004) Genotype and phenotype patterns of human immunodeficiency virus type 1 resistance to enfuvirtide during long-term treatment. *Antimicrob Agents Chemother* 48: 3253–3259

Entry Inhibitors in HIV Therapy
Edited by Jacqueline D. Reeves and Cynthia A. Derdeyn

Enfuvirtide: from basic science to FDA approval

Michael L. Greenberg

Trimeris Inc., 3500 Paramount Parkway, Morrisville, NC 27560, USA

Introduction

In the mid 1990s, the introduction of protease inhibitors (PIs) and non-nucleoside reverse transcriptase inhibitors (RTIs) enabled combinations of antiretroviral agents to be used in therapeutic regimens to treat HIV-infected patients. Such combination therapy often resulted in potent suppression of viral RNA and was generally accompanied by dramatically improved clinical outcomes for patients with HIV infection [1]. The potent suppression of HIV RNA led to such combinations of antiretroviral agents being referred to as highly active antiretroviral therapy (HAART), and this approach to treatment of HIV disease has resulted in striking and sustained decreases in AIDS-related death since its introduction in 1996 [2]. Prior to the 2003 approval and introduction of Enfuvirtide (Fuzeon™, formerly known as T-20), the first fusion inhibitor, only three classes of antiretrovirals were available for the treatment of HIV: the RTIs – either nucleoside or non-nucleoside – and PIs. Although combinations of these agents brought about dramatic improvements in HIV therapy, the limitations of therapeutic regimens based solely on RTIs and PIs were already evident and problematic when enfuvirtide entered clinical development. These limitations included adverse effects associated with treatment [3, 4], significant drug–drug interactions [5] and the selection of drug-resistant viruses with extensive intra-class cross-resistance [6]. Consequently, an unmet medical need existed (and still remains) for new classes of antiretroviral agents with both improved safety and tolerability profiles, and which circumvent the problems associated with intra-class cross-resistance by acting on alternative targets to inhibit viral replication. Since the introduction of PIs in 1995 and non-nucleoside RTIs in 1996, the occurrence of intra-class resistance has often limited the impact of newly introduced agents in the existing three classes of antiretroviral agents. However, improvements in patient management such as the earlier identification of virological failure, the use of resistance testing to optimize treatment regimens, and changes in the way in which antiretrovirals are used, such as PI boosting, and the more recent introduction of new entrants into those existing classes with enhanced barriers to resistance, such as the PIs Tipranavir and Darunavir, have provided welcome improvements in overall treatment durability. The introduction of enfuvirtide represented the first new class of antiretroviral agent available for the treatment of HIV-1 in 7 years and extended those improvements in treatment outcomes and durability even further. The clinical success of enfuvirtide fostered work on the process of HIV entry as a target for developing new therapeutics and opened up the entire field of entry inhibitors for treatment of HIV disease. The process of HIV entry offers many potential opportunities for intervention as evidenced throughout this book. HIV entry is a particularly attractive target for antiretroviral development since drugs targeting the entry process would not be expected to show cross-resistance to previous classes of antiretrovirals. This chapter reviews the science leading to the discovery and development of enfuvirtide and the clinical benefits that have been achieved by inclusion of enfuvirtide into combination therapeutic regimens for treatment-experienced patients.

Early investigation of the HIV envelope as a vaccine target identified peptide inhibitors of HIV infection

Initial efforts in the search for a vaccine to prevent HIV infection centered on the HIV envelope glycoprotein gp160 as a target for neutralizing antibodies [7–10]. The HIV envelope glycoprotein is a type I integral membrane protein translated as a polyprotein that is subsequently cleaved to yield the gp120 surface subunit and the gp41 transmembrane subunit, which associate as non-covalently bound oligomeric trimers on the surface of virions [11–15]. The process of HIV binding and entry into $CD4^+$ target cells involves several discrete and discernable steps leading to a series of conformational changes in both envelope subunits. The first step is the binding of the virus envelope via the gp120 subunit to the CD4 receptor on the target cell surface. This interaction leads to conformational changes in the gp120 subunit that exposes and creates a binding site for a chemokine coreceptor. The major coreceptors for HIV are the CCR5 and CXCR4 coreceptor, and the presence of one or both of these coreceptors on $CD4^+$ cells is required for virus entry [16–18]. The engagement of receptor and coreceptor (either CCR5 or CXCR4) by gp120 leads to a common set of structural rearrangements within the gp41 transmembrane subunit that are required for the fusion of the virus and target cell membranes [12]. These rearrangements in gp41 involve two heptad repeat regions in the ectodomain referred to as heptad repeat 1 (HR1, proximal to the N terminus and the fusion peptide region) and heptad repeat 2 (HR2, proximal to the C terminal region of the ectodomain near the membrane spanning region of the virus transmembrane protein). The two HR regions of the gp41 ectodomain are separated by a hinge region containing two cysteine residues that are able to form a disulfide-bonded loop.

The model for the native form of the envelope trimer envisions that each gp41 molecule is held in a high-energy conformation with the fusion peptide folded towards the viral membrane [19]. The fusion process involves the insertion of the N-terminal fusion peptide region of gp41 into the target cell membrane, the formation of a trimeric coiled-coil structure formed by the HR1 region and the binding of the HR2 region along the groves of the trimeric HR1 structure to form a stable six-helix bundle hairpin structure [19, 20] that brings the viral and cell membranes into close proximity for the fusion event. Aggregation of multiple activated envelope trimers leads to the creation of a fusion pore that permits the virus to enter the target cell [19, 21, 22]. During this structural rearrangement of gp41 prior to formation of the stable six-helix bundle, an intermediate pre-hairpin structure is formed, exposing the N-terminal coiled-coil region of gp41 [22, 23]. This process has been compared with the spring-loaded mechanism proposed for the action of the influenza HA2 protein [24, 25].

Epitope mapping studies conducted at Duke University Medical Center in the laboratory of Thomas Matthews employed a variety of synthetic peptides derived from both the gp120 and gp41 subunits of the envelope protein as part of efforts to identify potential targets for neutralizing antibodies. Such antibody responses and the peptide epitopes studied were often evaluated in a T cell-based cell-cell fusion assay [7]. An unexpected finding occurred when a synthetic peptide termed DP-107, derived from the HR1 region of gp41, was examined in the fusion assay that took this work in a decidedly different direction. DP-107 was found to block cell-cell fusion and virus infection [26]. Further studies demonstrated that the peptide possessed a very stable helical structure and that this structure was important for its biological activity [27]. Additional studies examined synthetic peptides from the C-terminal region of the gp41 ectodomain as these sequences were also predicted to adopt an alpha-helical structure. A 36-amino acid peptide from this region called DP-178 (later referred to as T-20 and subsequently enfuvirtide) blocked HIV fusion at concentrations approximately 100 times lower than DP-107 [28, 29] (Fig. 1). DP-178 was derived from amino acid residues 643–678 of HIV-1LAI gp160 and contains the epitope for the neutralizing monoclonal antibody 2F5 [30].

Mechanism of inhibition of the HIV fusion process by enfuvirtide and other HR2 peptides

A number of synthetic peptides that mimic the HR2 region of gp41 have subsequently been shown to exhibit significant antiviral activity *in vitro* and *in vivo*. A substantial body of *in vitro* and *in vivo* studies suggests that these peptides most probably act by competitively binding to the hydrophobic grooves of the HR1 trimer, thus preventing the formation of the hairpin structure and subsequent membrane fusion events. Enfuvirtide is thought to inhibit the fusion process by binding to HR1 and blocking its interaction with HR2 as described above [11, 31, 32]. Initial insights into the mechanism of enfuvirtide action were gleaned from studies with recombinant soluble gp41 fusion peptide constructs containing various portions of the ectodomain [31]. When a

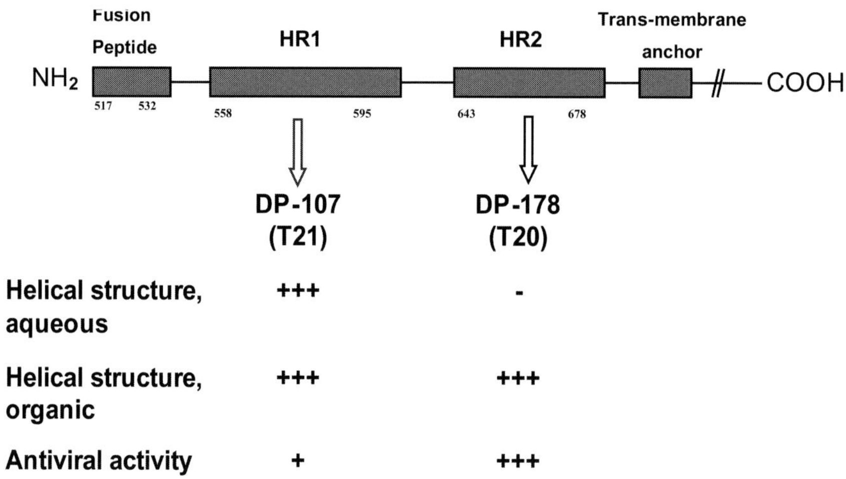

Figure 1. Peptide models of heptad repeats 1 and 2 (HR1 and HR2) in ectodomain of HIV-1 gp41 display antiviral activity. Schematic of the HIV gp41 ectodomain highlighting the locations and amino acid numbers corresponding to fusion peptide, HR1 and HR2 regions. The helical structures and antiviral activity observed with peptides DP-107 and DP-178 are indicated.

wild-type construct containing both HR regions was tested it was devoid of antiviral activity. However, when constructs were tested that contained either a mutation in the HR1 region that destroyed its helical structure or were deleted for the HR1 region, antiviral activity was found that was nearly identical to that observed with enfuvirtide (i.e., DP-178) [31]. These results (along with others) were taken to suggest that the HR1 and HR2 regions likely associated in the wild-type construct, thereby sequestering the enfuvirtide-containing region and rendering it unavailable. The mutations that either deleted the HR1 region or destroyed its helical structure probably freed the enfuvirtide-containing region and allowed it to inhibit virus infection [11, 31]. These notions were supported by subsequent crystal structures of the six-helix bundle showing the association of the HR1 and HR2 region [19, 20]. Later studies solidified these notions by demonstrating that the pre-hairpin intermediate is the target for enfuvirtide binding [32].

Enfuvirtide was shown to inhibit fusion mediated by the envelope of various strains of HIV-1 with EC_{50} values in the low nanomolar range. The activity of enfuvirtide against HIV-2 envelope-mediated cell fusion was about 1000-fold lower, indicating its selectivity for HIV-1. These results were confirmed by other investigators using various combinations of cell types and reporting EC_{50} values within a similar range [33–35]. In some *in vitro* studies, higher concentrations of enfuvirtide were required to block infection with cell-free virus than to inhibit cell-cell fusion. For example, Wild et al. reported enfuvirtide EC_{50} values for inhibition of cell-free infection with the laboratory isolate HIV-1_{LAI} of 90 ng/ml (20 nM) in CEM cells (*versus* 1.5 ng/ml to inhibit cell-cell fusion) and 1100 ng/ml (240 nM) in peripheral blood mononuclear cells [29]. The basis for these apparent differences in the inhibitory activity of enfuvirtide for cell-cell fusion *versus* infection by cell-free virus is not clear. They could reflect true differences in inhibitory activities of enfuvirtide or perhaps they simply demonstrate and remind us of the differences in inhibitory values that can be obtained with different assay systems. In either event, these studies illustrated a distinguishing feature of enfuvirtide derived from its mechanism of action – the ability to inhibit two modes of virus transmission – infection via cell-free virus and infection via cell-to-cell transmission.

Preclinical data had demonstrated that enfuvirtide possessed potent inhibitory activity *in vitro* and its mechanism of action suggested that it would be active against viruses that had acquired resistance to the more traditional RTIs and PIs. Nevertheless, because of its peptidic nature, enfuvirtide faced several challenges as a potential therapeutic (Tab. 1). Of concern was the lack of oral bioavailability, the notion that a peptide would have very poor pharmacokinetic properties and the potential for either an immunogenic response to the peptide or that pre-existing cross-reactive antibodies in HIV-1-infected patients would either neutralize enfuvirtide activity or facilitate clearance of enfuvirtide from the circulation (Tab. 1). However, preclinical animal studies in rodent and monkey models demonstrated that enfuvirtide possessed sufficient pharmacokinetic properties to be considered further as a potential therapeutic (Tab. 2). Importantly, enfuvirtide exhibited potent antiviral activity *in vivo* in the Hu-PBMC SCID mouse model and antibodies to enfuvirtide did not neutralize its antiviral activity *in vitro*.

Table 1. Challenges for peptides as therapeutics

1.	Not orally bioavailable
2.	Poor pharmacokinetic characteristics
3.	Can be antigenic – interference by antibody
4.	Complex and difficult to manufacture

Early clinical studies with enfuvirtide

The first proof of concept came in a Phase I/II 14-day trial (TRI-001). The trial employed intravenous infusion of enfuvirtide as monotherapy with doses ranging from 3 to 100 mg administered twice daily. In patients receiving the highest dose, the median plasma viral load reduction was 1.96 $\log_{10}$ copies of HIV RNA per ml and a clear dose response was evident [36]. A follow-up Phase I/II trial explored subcutaneous infusion (doses of 12.5–100 mg enfuvirtide/day) and twice daily subcutaneous injection (45 or 90 mg enfuvirtide bid). Dose-dependent decreases in plasma viral load were found with the maximum decrease observed for the 90 mg bid group [37]. Administration of enfuvirtide by twice daily subcutaneous injections yielded predictable pharmacokinetics and maintained trough levels well above the *in vitro* IC_{50} values. Subsequent Phase II studies demonstrated the longer-term safety profile and activity of enfuvirtide-containing regimens when the drug was administered by the subcutaneous injection and explored the dose-response relationship over 48 weeks of therapy. These studies demonstrated that injection by the subcutaneous route was well tolerated, with no dose-related toxicities [38, 39]. In addition, they yielded promising data on $CD4^+$ T cell gains during therapy with enfuvirtide-containing regimens in treatment-experienced patients and identified a higher strength carbonate buffer formulation of enfuvirtide for evaluation as a twice daily dose of 90 mg for Phase III trials [40].

Pharmacokinetics

The pharmacokinetics of enfuvirtide were examined in an open label, single-dose cross-over study design [41]. Bioavailability after a single 90 mg subcutaneous injection was high (84.3%) compared to a similar intravenous dose. The apparent plasma half-life for enfuvirtide was 3.8 h and the apparent systemic clearance was 1.68 l/h (Tab. 2). The Cmax, Cmin and AUC exhibited dose dependency, while the half-life and clearance were dose independent. Absorption of enfuvirtide from different sites of injection (abdomen, arm and thigh) was comparable [42]. Enfuvirtide may be excreted unchanged, deamidated and excreted, or catabolized to its constituent amino acids followed by recycling of these amino acids in the body pool rather than being metabolized by the cytochrome systems. Because of its anticipated catabolic route, there is a low potential for enfuvirtide to be affected by or to affect other drugs commonly used in HIV-infected patients. This assumption has been investigated in several studies including a "cocktail" study (assessing effects on caffeine for CYP1A2, chlorzoxone for CYP2E1, dapsone for CYP3A4, debrisoquine for CYP2D6 and mephenytoin for CYP2C19) [43] and studies investigating effects of rifampicin, ritonavir and the combination of saquinavir with ritonavir [44, 45]. There were no clinically meaningful effects of enfuvirtide on the drugs studied or of the drugs studied on enfuvirtide. Thus, these studies confirmed the low potential for drug-drug interactions to be a significant issue with enfuvirtide administration. More recently, there has been a report of high-

Table 2. T-20 possesses drug like properties in preclinical studies

- T1/2 = 2.5 h in plasma
- Bioavailability 60–80% as subcutaneous injection
- Readily reaches lymphatic system
- Activity in Hu-PBMC SCID model of HIV-1 infection
- Antibodies to T-20 do not neutralize activity

er concentrations of tipranavir and ritonavir trough concentrations in patients receiving enfuvirtide [46]. The mechanism underlying this unexpected interaction is unclear and requires further examination.

Pivotal Phase III efficacy trials

The long-term safety and efficacy of enfuvirtide was demonstrated in two large Phase III trials termed TORO (T-20 *versus* Optimized Regimen Only) 1 and 2 [47, 48]. The trial designs were very similar and enrolled similar patient populations. The designs differed only in the minimum length of previous experience for antiretroviral agents from each of the three available classes (6 months for TORO 1 *versus* 3 months for TORO 2), and the number of previous PIs that patients must have received (two for TORO 1 *versus* one for TORO 2). The studies enrolled approximately 1000 highly treatment-experienced patients who had either been exposed to or had documented resistance to agents from each of the three approved classes of antiretrovirals. Patients received enfuvirtide in combination with an antiretroviral regimen optimized based upon drug resistance testing and patient history or the optimized regimen alone. The optimized regimen contained from three to five drugs. Patients were randomized in a 2:1 ratio to receive either the enfuvirtide-containing regimen or the optimized regimen alone. One of the novel aspects of the TORO trial designs was an early "escape" for patients in the optimized regimen alone group if they experienced confirmed virological failure during the trials. Such patients could revise or reoptimize their regimen based upon new resistance testing and add enfuvirtide. The primary efficacy parameter in both trials was the change from baseline in plasma HIV-1 RNA.

In both trials, at 24 weeks, the group receiving enfuvirtide in combination with an optimized regimen exhibited significantly greater declines in viral RNA than the group treated with the optimized regimen alone, approaching almost a one $\log_{10}$ difference between the groups [47, 48]. Importantly, these studies also demonstrated improved $CD4^+$ T cell responses for the enfuvirtide-treatment groups. Due to similarities in the trial designs and patient populations, a pooled analysis of both studies was undertaken to provide a more robust assessment of the treatment effect and enable subgroup analysis to be conducted with sufficient power. The pooled analysis also showed that the reduction in RNA and the increase in $CD4^+$ T cell counts were significantly greater for the enfuvirtide-treatment group at both 24 and 48 weeks. The intent to treat analysis demonstrated that the enfuvirtide-treatment group exhibited an additional 0.84 $\log_{10}$ decrease in viral load at 24 weeks, and this difference in treatment effect was maintained throughout 48 weeks of therapy (0.85 log10 difference at week 48) [49]. These findings demonstrated the durable nature of the virological response to enfuvirtide therapy achieved in these trials. Enfuvirtide therapy also led to a doubling of the $CD4^+$ T cell response in comparison to the optimized regimen alone group at both 24 and 48 weeks, with gains of 71 and 91 cells, respectively, for the enfuvirtide-treatment group [49]. The immunological response was particularly noteworthy given the advanced nature of the enrolled patient population. The efficacy results described briefly above combined with the favorable safety profile of Fuzeon in the clinical trials [50] formed the basis for its accelerated approval in 2003 and subsequent traditional approval in 2004, establishing Fuzeon as the first fusion inhibitor for treatment of HIV. Local injection site reactions were the most common adverse event associated with enfuvirtide treatment, with approximately 98% of patients on the enfuvirtide-containing arms in the TORO trials having experienced an injection site reaction. Injection site reactions are perhaps the most significant adverse event considered by physicians and patients when evaluating whether to initiate Fuzeon therapy.

Clinical results obtained in the Phase III TORO studies of Fuzeon along with more recent studies of the newly approved PIs tipranavir and darunavir have demonstrated the clinical benefits that can be achieved in treatment-experienced HIV-infected patients when the fusion inhibitor Fuzeon is combined with an active boosted PI [51, 52]. These striking clinical results in treatment-experienced patients were important considerations in the 2006 revision of recommendations from the International AIDS Society-USA for treatment for adult HIV infection in the setting of multiple regimen failure [53]. That panel now recommends when two or more potent active drugs are identified, the goal of therapy should be suppression of viral load to below 50 copies/ml, even for highly treatment-experienced patients. The recommendations discuss how best to achieve that goal and add that inclusion of enfuvirtide is often needed to achieve this in heavily treatment-experienced patients [53].

Enfuvirtide activity and mechanism of drug resistance

Susceptibility of fusion inhibitor-naïve isolates and correlation with clinical response

Much has been learned from both clinical- and laboratory-based studies about the activity of enfuvirtide and the ways that HIV-1 can become resistant to its antiviral activity. Fusion inhibitor-naïve viruses exhibit a broader range of susceptibility to inhibition *in vitro* by enfuvirtide than had been seen with either nucleoside RTIs or PIs tested against viruses naïve to those classes of inhibitors. *In vitro* studies with clinical isolates from patients participating in Phase II trials of enfuvirtide demonstrated that enfuvirtide exhibited a wide range (over 100-fold) of IC_{50}s for primary isolates with a geometric mean IC_{50} of approximately 20 ng/ml (4 nM) in a cMAGI-based assay [54]. For the Phase III TORO 1 and TORO 2 studies, susceptibility to enfuvirtide was assessed with the PhenoSense entry assay (Monogram Biosciences, South San Francisco, CA). At baseline, the range of IC_{50} values observed was approximately 1000-fold and log normally distributed around a geometric mean of

0.26 μg/ml [55]. Examination of the relationship between baseline IC_{50} and virological response found no difference in the response for patients with viruses whose susceptibility was more than 2 standard deviations outside of the geometric mean sensitivity. The range of *in vitro* susceptibility to inhibition by enfuvirtide extended from IC_{50} values near 10 ng/ml to over 7 μg/ml. Importantly, amongst these fusion inhibitor-naïve patients, lower *in vitro* baseline susceptibility was not associated with a significantly decreased virological response to enfuvirtide *in vivo* [55].

Determinants of enfuvirtide susceptibility and resistance

Enfuvirtide targets the HR1 region of the viral gp41 transmembrane protein. Initial *in vitro* studies selecting for escape variants [56], as well as clinical virology examination of samples from patients experiencing virological rebound in Phase I/II clinical studies [54, 57], suggested that gp41 amino acids 36–45 were involved in the development of resistance to enfuvirtide. A large body of data including those from the Phase III TORO studies [55] and a number of studies from additional groups [58] confirm that the initial development of resistance to enfuvirtide mostly involves changes in the gp41 HR1 target region and appears within amino acids 36–45. In the TORO studies, amino acid substitutions within this region were found in nearly 93% of patients meeting protocol-defined virological failure criteria, and occurred in almost all cases (98.8%) where susceptibility to enfuvirtide had decreased by fourfold or greater [55]. These findings provide compelling genetic evidence in support of enfuvirtide's mechanism of action and target. In clinical studies (mostly carried out in patients infected with Clade B HIV viruses) the most commonly observed resistance mutations have been V38A, N43D, G36D, and Q40H.

Genotypic studies of viruses from fusion inhibitor-naïve individuals [59, 60] as well as HIV envelope sequences deposited in the Los Alamos Database (http://hiv-web.lanl.gov) indicate that HIV-1 gp41 amino acids 36–45 are highly conserved (particularly in clade B), with the exception of codon 42, which displays moderate polymorphism. Genotypic analyses of virus derived from fusion inhibitor-naive patients enrolled in Phase II and Phase III clinical trials of enfuvirtide confirm this observation [54, 55], and indicate that the N42S polymorphism, which does not negatively impact susceptibility to enfuvirtide, occurs at a frequency of approximately 16%.

In spite of the conserved nature of the HR1 amino acids 36–45 in fusion inhibitor-naive viruses and the absence of mutations known to confer reduced susceptibility to enfuvirtide [55, 60], as mentioned above clinical isolates from fusion inhibitor-naive patients display a broad range of susceptibilities to enfuvirtide. The basis for this phenotypic heterogeneity is not entirely clear but appears to be related to polymorphisms in the HR2 region of gp41 and other regions of gp41 outside of the target region for enfuvirtide [58, 61, 62]. In addition, V3 loop determinants of coreceptor specificity within the viral gp120 envelope protein and coreceptor density can modulate *in vitro* sensitivity to enfuvirtide [63–65]. Dual/mixed tropic viruses (encompassing the vast majority of clinical isolates with the ability to infect via the CXCR4 coreceptor [66]) display lower *in vitro* susceptibilities to enfuvirtide than R5 viruses [55, 63, 64]. The significance of the differential *in vitro* sensitivity to enfuvirtide exhibited by dual/mixed tropic viruses *versus* R5 tropic viruses is unclear since clinical studies have demonstrated that patients harboring both types of viruses respond equally well to enfuvirtide-containing regimens [55, 66]. Nevertheless, it is possible that this differential susceptibility is related to the shift in virus tropism from dual/mixed tropic to R5 tropic that was observed at viral failure in the TORO studies [66].

Enfuvirtide resistance mutations impact viral "fitness"

The conserved nature of the gp41 HR1 sequence and its essential role in virus entry suggests that mutations in this region may have a negative impact on the replicative capacity or "fitness" of the virus. Lu et al. [67] were among the first to demonstrate fitness costs associated with enfuvirtide-resistance mutations. These investigators employed growth competition assays with NL4-3 clones carrying various HR1 mutations, as well as baseline and on-treatment patient envelopes to measure the fitness costs associated with introduction of enfuvirtide-resistance mutations into various HIV-1 envelopes. Their work suggested a relationship between loss of enfuvirtide susceptibility and deficits in virus replication [67]. In contrast to the results from Lu et al., some investigators have failed to find a consistent effect of enfuvirtide mutations on viral replication capacity even thought they do find that enfuvirtide-resistance mutations rapidly disappear from the plasma virus population when the drug is withdrawn [68]. The rapid disappearance of enfuvirtide-resistance mutations following discontinuation of enfuvirtide therapy in patients experiencing virological failure appears to be a consistent finding [69], which strongly implies an *in vivo* fitness cost for enfuvirtide-resistance mutations.

New insights into enfuvirtide resistance and virus pathogenesis

In the past several years there has been an increasing appreciation that ongoing viral replication in the presence of long-term enfuvirtide therapy is associated with evolution of enfuvirtide resistance and can involve changes in the HR2 region of gp41 in addition to those occurring in HR1 [68, 70, 71]. While most studies suggest the impact of these secondary or compensatory mutations in HR2 on enfuvirtide sensitivity *in vitro* appear to be more modest than the primary HR1 resistance mutations [62, 70–72], their effects on virus replication are not well characterized. Recent crystallographic studies with peptide models have shed light on the molecular and structural basis for decreased enfuvirtide activity for viruses bearing the N43D HR1 mutation, and provide a framework for understanding the preferential occurrence of this mutation in viruses that have the E137K polymorphism in HR2 [73]. These studies suggest that there is much to learn about the structural consequences of HR1 resistance mutations along with the altered interactions of these substituted amino acids with HR2 residues and the consequences on six-helix bundle formation. Further insights into these aspects may provide a greater understanding of the mechanisms and impact of enfuvirtide resistance mutations on six-helix bundle structure, virus replication and pathogenicity.

In regard to the latter points, Reeves and colleagues [74] have shown that certain HR1 mutations (in the absence of HR2 mutations) alter envelope function, leading to decreased fusion kinetics and an increased sensitivity of virus to neutralization by a subset of neutralizing monoclonal antibodies and some HIV-1-positive sera. Baldwin and colleagues [75] have suggested that the combination of the V38A mutation with the N126K HR2 mutation can have harmful effects on the tightly regulated fusion process with adverse consequences for virus replication. Studies have indicated that the HIV-1 envelope is a major determinant of virus pathogenicity and $CD4^+$ T-cell depletion [76]. It currently remains unclear how the effects of enfuvirtide-resistance mutations on envelope function impact the response to therapy, although several studies suggest intriguing possibilities. Poveda et al. [71] were the first to report that some patients experiencing viral rebound and phenotypic resistance during enfuvirtide therapy continue to maintain elevations in $CD4^+$ T cells. Carlo Perno's group [77] examined virological and immunological responses in a larger group of 54 subjects where enfuvirtide was added to a failing background antiviral regimen. These investigators noted differential outcomes for patients based upon specific enfuvirtide-resistance mutations that arose during therapy with enfuvirtide. Thus, the group of patients who acquired the most common mutations at V38 (A or E) continued to experience $CD4^+$ T cell gains over 36 weeks despite virological rebound. This result was in contrast to the group of patients who developed the less common Q40H + L45M mutations and experienced a loss of CD4 cells over the 36 week period of the study. The results from the Perno's group were confirmed and extended by Melby and colleagues [78] who examined changes in $CD4^+$ T cell counts in patients experiencing virological failure in the TORO Phase III studies. Melby et al. [78] found that the group of patients failing with substitutions at position V38 exhibited continued $CD4^+$ T cell gains through 48 weeks of continued enfuvirtide therapy post-viral failure. In contrast (and similar to the results from the Perno group), they also found that patients developing the Q40H mutation experienced a loss of $CD4^+$ T cells during continued enfuvirtide therapy after viral failure. Both groups suggested that their findings were consistent with reduced pathogenicity for viruses bearing the V38 mutation. Whether these effects are due to alterations in virus fusion kinetics or efficiency is unknown. Some investigators have suggested the continued $CD4^+$ T cell gains may be due to decreases in immunological activation associated with enfuvirtide therapy [71, 79]. Other mechanisms that could be involved include decreases in gp41-induced apoptosis and bystander cell death brought about by enfuvirtide and/or enfuvirtide-resistance mutations [79–81]. Clearly more work is required to understand the mechanisms that contribute to the gains and/or maintenance of elevated CD4 cells in patients continuing enfuvirtide therapy in the presence of ongoing viral replication and enfuvirtide resistance.

Conclusions and future directions

The fusion inhibitor enfuvirtide acts by binding to the HR1 region of HIV gp41 transmembrane protein, and thereby interrupts a critical step in viral entry – formation of the six-helix bundle – providing a new therapeutic target to treat HIV infection. The Phase III TORO studies demonstrated the clinical efficacy and safety of enfuvirtide-based therapy in treatment-experienced patients. Enfuvirtide containing regimens that include an active boosted PI have resulted in previously unattainable virological and immunological responses in heavily treatment-experienced patients, leading to new treatment recommendations that bridge the gap in treatment goals for this group of patients with those of treatment-naïve patients. These clinical results demonstrate the benefit this new class of fusion inhibitors can bring to therapeutic regimens for highly treatment-experienced patients. A recent report of a study in treatment-experienced patients suggests that combining two new classes of antiviral agents, specifically enfuvirtide and the MK-0518 integrase inhibitor currently in development, can further extend these benefits such that by 24 weeks 90% or more of patients receiving these agents for the first time may achieve viral loads less than 400 copies/ml [82].

The availability of additional entry inhibitors that target different steps in the process raises the possibility that enfuvirtide could be combined with other entry inhibitors. Laboratory and clinical studies clearly implicate substitutions within HR1 at gp41 amino acids 36–45 in the development of resistance to enfuvirtide. Although the potential for cross-resistance between enfuvirtide and CD4 or coreceptor binding inhibitors has not been extensively explored, available data suggest that enfuvirtide-resistance mutations may have little if any influence on sensitivity to inhibitors targeting CD4 or coreceptor interactions [74]. Thus, the use of these entry inhibitors and enfuvirtide in combination or in sequence may be possible and advantageous. *In vitro* data suggest possible synergistic benefits to such combinations [65, 83], which will need validation through clinical trials.

Enfuvirtide was developed to address an unmet medical need for treatment-experienced patients with limited options remaining to suppress their virus. As new agents have become available to combine with enfuvirtide, the percentage of such patients that can become fully suppressed is approaching what can be achieved in first line regimens. These striking clinical results, combined with enfuvirtide's safety profile have led to continued work on development of a next generation fusion inhibitor with a significantly improved target profile. TRI-1144 is a lead candidate fusion inhibitor peptide in preclinical development that exhibits a more robust genetic barrier for development of resistance and importantly, physical and pharmacokinetic properties that may allow for once weekly dosing.

Acknowledgements
The author would like to acknowledge the many contributions of colleagues at Trimeris and Roche who contributed to the development of enfuvirtide and our understanding of how it works. I would like to give special thanks to my Trimeris colleagues Tom Melby, Ralph DeMasi, Dani Bolognesi and Thomas Matthews (deceased) and to my Roche colleagues Neil Buss, Mickey Salgo, Gabrielle Heilik-Snyder and Nick Cammack.

References

1 Palella FJ Jr, Delaney KM, Moorman AC, Loveless MO, Fuhrer J, Satten GA, Aschman DJ, Holmberg SD (1998) Declining morbidity and mortality among patients with advanced human immunodeficiency virus infection. HIV Outpatient Study Investigators. *N Engl J Med* 338: 853–860
2 Palella FJ, Jr, Baker RK, Moorman AC, Chmiel JS, Wood KC, Brooks JT, Holberg SD and HIV Outpatient Study Investigators (2006) Mortality in the highly active antiretroviral therapy era: changing causes of death and disease in the HIV outpatient study. *J Acquir Immune Defic Syndr* 43: 27–34
3 Fellay J, Boubaker K, Ledergerber B, Bernasconi E, Furrer H, Battegay M, Hirschel B, Vernazza P, Francioli P, Greub G et al. (2001) Prevalence of adverse events associated with potent antiretroviral treatment: Swiss HIV Cohort Study. *Lancet* 358: 1322–1327
4 Yeni PG, Hammer SM, Carpenter CC, Cooper DA, Fischl MA, Gatell JM, Gazzard BG, Hirsch MS, Jacobsen DM, Katzenstein DA et al. (2002) Antiretroviral treatment for adult HIV infection in 2002: updated recommendations of the International AIDS Society-USA Panel. *JAMA* 288: 222–235
5 Kosel BW, Aweeka F (2000) Drug interactions of antiretroviral agents. *AIDS Clin Rev* 193–227
6 Shafer RW, Winters MA, Palmer S, Merigan TC (1998) Multiple concurrent reverse transcriptase and protease mutations and multidrug resistance of HIV-1 isolates from heavily treated patients. *Ann Intern Med* 128: 906–911
7 Javaherian K, Langlois AJ, McDanal C, Ross KL, Eckler LI, Jellis CL, Profy AT, Rusche JR, Bolognesi DP, Putney SD, Matthews TJ (1989) Principal neutralizing domain of the human immunodeficiency virus type 1 envelope protein. *Proc Natl Acad Sci USA* 86: 6768–6772
8 Berman PW, Gregory TJ, Riddle L, Nakamura GR, Champe MA, Porter JP, Wurm FM, Hershberg RD, Cobb EK, Eichberg JW (1990) Protection of chimpanzees from infection by HIV-1 after vaccination with recombinant glycoprotein gp120 but not gp160. *Nature* 345: 622–625
9 Emini EA, Nara PL, Schleif WA, Lewis JA, Davide JP, Lee DR, Kessler J, Conley S, Matsuhita S, Putney SD et al. (1990) Antibody-mediated *in vitro* neutralization of human immunodeficiency virus type 1 abolishes infectivity for chimpanzees. *J Virol* 64: 3674–3678
10 Girard M, Kieny M-P, Pinter A, Barre-Sinoussi F, Nara P, Kolbe H, Kusumi K, Chaput A, Reinhart T, Muchmore E et al. (1991) Immunization of chimpanzees confers protection against challenge with human immunodeficiency virus. *Proc Natl Acad Sci USA* 88: 542–546
11 Matthews TJ, Wild C, Chen N, Bolognesi D, Greenberg ML (1994) Structural rearrangements in the transmembrane glycoprotein after receptor binding. *Immunol Rev* 140: 93–104
12 Eckert DM, Kim PS (2001) Mechanisms of viral membrane fusion and its inhibition. *Annu Rev Biochem* 70: 777–810
13 Einfeld D, Hunter E (1988) Oligomeric structure of a prototype retrovirus glycoprotein. *Proc Natl Acad Sci USA* 85: 8688–8692
14 Pinter A, Honnen WJ, Tilley SA, Bona C, Zaghouani H, Gorny MK, Zolla-Pazner S (1989) Oligomeric structure of gp41, the transmembrane protein of human immunodeficiency virus type

1. *J Virol* 63: 2674–2679
15 Wyatt R, Sodroski J (1998) The HIV-1 envelope glycoproteins: fusogens, antigens, and immunogens. *Science* 280: 1884–1888
16 Berson JF, Long D, Doranz BJ, Rucker J, Jirik FR, Doms RW (1996) A seven-transmembrane domain receptor involved in fusion and entry of T-cell-tropic human immunodeficiency virus type 1 strains. *J Virol* 70: 6288–6295
17 Deng H, Liu R, Ellmeier W, Choe S, Unutmaz D, Burkhart M, Di Marzio P, Marmon S, Sutton RE, Hill CM et al. (1996) Identification of a major co-receptor for primary isolates of HIV-1. *Nature* 381: 661–666
18 Feng Y, Broder CC, Kennedy PE, Berger EA (1996) HIV-1 entry cofactor: functional cDNA cloning of a seven-transmembrane, G protein-coupled receptor. *Science* 272: 872–877
19 Weissenhorn W, Dessen A, Harrison SC, Skehel JJ, Wiley DC (1997) Atomic structure of the ectodomain from HIV-1 gp41. *Nature* 387: 426–430
20 Chan DC, Fass D, Berger JM, Kim PS (1997) Core structure of gp41 from the HIV envelope glycoprotein. *Cell* 89: 263–273
21 Furuta RA, Wild CT, Weng Y, Weiss CD (1998) Capture of an early fusion-active conformation of HIV-1 gp41. *Nat Struct Biol* 5: 276–279
22 Melikyan GB, Markosyan RM, Hemmati H, Delmedico MK, Lambert DM, Cohen FS (2000) Evidence that the transition of HIV-1 gp41 into a six-helix bundle, not the bundle configuration, induces membrane fusion. *J Cell Biol* 151: 413–423
23 Munoz-Barroso I, Durell S, Sakaguchi K, Appella E, Blumenthal R (1998) Dilation of the human immunodeficiency virus-1 envelope glycoprotein fusion pore revealed by the inhibitory action of a synthetic peptide from gp41. *J Cell Biol* 140: 315–323
24 Bullough PA, Hughson FM, Skehel JJ, Wiley DC (1994) Structure of influenza haemagglutinin at the pH of membrane fusion. *Nature* 371: 37–43
25 Carr CM, Kim PS (1993) A spring-loaded mechanism for the conformational change of influenza hemagglutinin. *Cell* 73: 823–832
26 Wild C, Oas T, McDanal C, Bolognesi D, Matthews T (1992) A synthetic peptide inhibitor of human immunodeficiency virus replication: correlation between solution structure and viral inhibition. *Proc Natl Acad Sci USA* 89: 10537–10541
27 Wild C, Dubay JW, Greenwell T, Baird T, Oas TG, McDanal C, Hunter E, Matthews T (1994) Propensity for a leucine zipper-like domain of human immunodeficiency virus type 1 gp41 to form oligomers correlates with a role in virus-induced fusion rather than assembly of the glycoprotein complex. *Proc Natl Acad Sci USA* 91: 12676–12680
28 Wild C, Greenwell T, Matthews T (1993) A synthetic peptide from HIV-1 gp41 is a potent inhibitor of virus-mediated cell-cell fusion. *AIDS Res Hum Retroviruses* 9: 1051–1053
29 Wild CT, Shugars DC, Greenwell TK, McDanal CB, Matthews TJ (1994) Peptides corresponding to a predictive alpha-helical domain of human immunodeficiency virus type 1 gp41 are potent inhibitors of virus infection. *Proc Natl Acad Sci USA* 91: 9770–9774
30 Zwick MB (2005) The membrane-proximal external region of HIV-1 gp41: a vaccine target worth exploring. *AIDS* 19: 1725–1737
31 Chen CH, Matthews TJ, McDanal CB, Bolognesi DP, Greenberg ML (1995) A molecular clasp in the human immunodeficiency virus (HIV) type 1 TM protein determines the anti-HIV activity of gp41 derivatives: implication for viral fusion. *J Virol* 69: 3771–3777
32 Kliger Y, Gallo SA, Peisajovich SG, Munoz-Barroso I, Avkin S, Blumenthal R, Shai Y (2001) Mode of action of an antiviral peptide from HIV-1. Inhibition at a post-lipid mixing stage. *J Biol Chem* 276: 1391–1397
33 Lawless MK, Barney S, Guthrie KI, Bucy TB, Petteway SR, Merutka (1996) HIV-1 membrane fusion mechanism: structural studies of the interactions between biologically-active peptides from gp41. *Biochemistry* 35:13697–13708
34 Nagashima KA, Thompson DA, Rosenfield SI, Maddon PJ, Dragic T, Olson WC (2001) Human immunodeficiency virus type 1 entry inhibitors PRO 542 and T-20 are potently synergistic in blocking virus-cell and cell-cell fusion. *J Infect Dis* 183: 1121–1125
35 Pine PS, Weaver JL, Oravecz T, Pall M, Ussery M, Aszalos A (1998) A semiautomated fluorescence-based cell-to-cell fusion assay for gp120-gp41 and CD4 expressing cells. *Exp Cell Res* 240: 49–57
36 Kilby JM, Hopkins S, Venetta TM, Di Massimo B, Cloud GA, Lee JY, Alldredge L, Hunter E, Lambert D, Bolognesi D et al. (1998) Potent suppression of HIV-1 replication in humans by T-20,

a peptide inhibitor of gp41-mediated virus entry. *Nat Med* 4: 1302–1307
37 Kilby JM, Lalezari JP, Eron JJ, Carlson M, Cohen C, Arduino RC, Goodgame JC, Gallant JE, Volbergding P, Murphy RL et al. (2002) The safety, plasma pharmacokinetics, and antiviral activity of subcutaneous enfuvirtide (T-20), a peptide inhibitor of gp41-mediated virus fusion, in HIV-infected adults. *AIDS Res Hum Retroviruses* 18: 685–693
38 Lalezari JP, Eron JJ, Carlson M, Cohen C, De Jesus E, Arduino RC, Gallant JE, Volberding P, Murphy RL, Valentine F et al. (2003) A phase II clinical study of the long-term safety and antiviral activity of enfuvirtide based antiretroviral therapy. *AIDS* 17: 691–698
39 Lalezari JP, De Jesus E, Northfelt DW, Richmond G, Wolfe P, Haubrich R, Henry D, Powderly W, Becker S, Thompson M et al. (2003) A controlled Phase II trial assessing three doses of enfuvirtide (T-20) in combination with abacavir, amprenavir, ritonavir and efavirenz in non-nucleoside reverse transcriptase inhibitor-naïve HIV-infected adults. *Antiviral Ther* 8: 279–287
40 Wheeler DA, Lalezari JP, Kilby JM, Wheat J, Delehanty J, DeMasi R, Patel I, Salgo M. (2004) Safety, tolerability, and plasma pharmacokinetics of high-strength formulations of enfuvirtide (T-20) in treatment-experienced HIV-1-infected patients. *J Clin Virol* 30: 183–190
41 Zhang X, Nieforth K, Lang J, Rouzier-Panis R, Reynes J, Dorr A, Kolis S, Stiles MR, Kinchelow T, Patel IH (2002) Pharmacokinetics of plasma enfuvirtide after subcutaneous administration to patients with human immunodeficiency virus: Inverse Gaussian density absorption and 2-compartment disposition. *Clin Pharmacol Ther* 72: 10–19
42 Lalezari JP, Patel IH, Zhang X, Dorr A, Hawker N, Siddique Z, Kolis SJ, Kinchelow T (2003) Influence of subcutaneous injection site on the steady-state pharmacokinetics of enfuvirtide (T-20) in HIV-1-infected patients. *J Clin Virol* 28: 217–222
43 Zhang X, Lalezari JP, Badley AD, Dorr A, Kolis SJ, Kinchelow T, Patel IH (2004) Assessment of drug-drug interaction potential of enfuvirtide in human immunodeficiency virus type 1-infected patients. *Clin Pharmacol Ther* 75: 558–568
44 Boyd MA, Zhang X, Dorr A, Ruxrungtham K, Kolis S, Nieforth K, Kinchelow T, Buss N, Patel IH (2003) Lack of enzyme-inducing effect of rifampicin on the pharmacokinetics of enfuvirtide. *J Clin Pharmacol* 43: 1382–1391
45 Ruxrungtham K, Boyd M, Bellibas SE, Zhang X, Dorr A, Kolis S, Kinchelow T, Buss N, Patel IH (2004) Lack of interaction between enfuvirtide and ritonavir or ritonavir-boosted saquinavir in HIV-1-infected patients. *J Clin Pharmacol* 44: 793–803
46 de Requena DG, Calcagno A, Bonora S, Ladetto L, D'Avolio A, Sciandra M, Siccardi M, Bargiacchi O, Sinicco A, Perri GD (2006) Unexpected drug–drug interaction between tipranavir/ritonavir and enfuvirtide. *AIDS* 20: 1977–1979
47 Lalezari JP, Henry K, O'Hearn M, Montaner JS, Piliero PJ, Trottier B, Walmsley S, Cohen C, Kuritzkes DR, Eron JJ et al. (2003) Enfuvirtide, an HIV-1 fusion inhibitor, for drug-resistant HIV infection in North and South America. *N Engl J Med* 348: 2175–2185
48 Lazzarin A, Clotet B, Cooper D, Reynes J, Arasteh K, Nelson M, Katlama C, Stellbrink HJ, Delfraissy JF, Lange J et al. (2003) Efficacy of enfuvirtide in patients infected with drug-resistant human immunodeficiency virus type 1 in Europe and Australia. *N Engl J Med* 348: 2186–2195
49 Nelson M, Arasteh K, Clotet B, Cooper DA, Henry K, Katlama C, Lalezari JP, Lazzarin A, Montaner JS, O'Hearn M et al. (2005) Durable efficacy of enfuvirtide over 48 weeks in heavily treatment-experienced HIV-1-infected patients in the T-20 *versus* optimized background regimen only 1 and 2 clinical trials. *J Acquir Immune Defic Syndr* 40: 404–412
50 Trottier B, Walmsley S, Reynes J, Piliero P, O'Hearn M, Nelson M, Montaner J, Lazzarin A, Lalezari J, Katlama C et al. (2005) Safety of enfuvirtide in combination with an optimized background of antiretrovirals in treatment-experienced HIV-1-infected adults over 48 weeks. *J Acquir Immune Defic Syndr* 40: 413–421
51 Katlama C, Walmsley S, Hicks C, Cahn P, Neubacher D, Villacian J, and The Resist Investigators (2006) Tipranavir achieves twice the rate of treatment responses and prolongs durability of response *versus* comparator PI in antiretroviral experienced patients, independent of baseline $CD4^+$ cell count or viral load: 48 week Resist 1 and 2 combined analysis. In: Program and Abstracts of the 13th Conference on Retroviruses and Opportunistic Infections; February 5–8, 2006; Dever, CO, abstr. no. 520
52 Pozniak A, Saag M, Bellos N, Chiliade P, Grinsztejn B, Molina JM (2006) Efficacy of TMC114/r in treatment-experienced HIV patients: factors influencing outcome in the pooled 24-week analysis of Power 1, 2 and 3. In: Program and Abstracts of the 12th Annual Conference of the British HIV Association; March 29–April 1, 2006; Brighton, England, abstr. no. 3

53 Hammer SM, Saag MS, Schecter M, Montaner JSG, Schooley RT, Montaner JS, Schooley RT, Jacobsen DM, Thompson MA, Carpenter CC, Fischl MA, Gazzard BG et al. (2006) Treatment for Adult HIV Infection 2006 Recommendations of the International AIDS Society-USA Panel. *J Am Med Assoc* 296: 827–843

54 Sista PR, Melby T, Davison D, Jin L, Mosier S, Mink M, Nelson EL, DeMasi R, Cammack N, Salgo MP et al. (2004) Characterization of determinants of genotypic and phenotypic resistance to enfuvirtide in baseline and on-treatment HIV-1 isolates. *AIDS* 18: 1787–1794

55 Melby T, Sista P, DeMasi R, Kirkland T, Roberts N, Salgo M, Heilek-Snyder G, Cammack N, Matthews TJ, Greenberg ML (2006) Characterization of envelope glycoprotein gp41 genotype and phenotypic susceptibility to enfuvirtide at baseline and on treatment in the phase III clinical trials TORO-1 and TORO-2. *AIDS Res Hum Retroviruses* 22: 375–385

56 Rimsky LT, Shugars DC, Matthews TJ (1998) Determinants of human immunodeficiency virus type 1 resistance to gp41-derived inhibitory peptides. *J Virol* 72: 986–993

57 Wei X, Decker JM, Liu H, Zhang Z, Arani RB, Kilby JM, Saag MS, Wu X, Shaw GM, Kappes JC (2002) Emergence of resistant human immunodeficiency virus type 1 in patients receiving fusion inhibitor (T-20) monotherapy. *Antimicrob Agents Chemother* 46: 1896–1905

58 Briz V, Poveda E, Soriano V (2006) HIV entry inhibitors: mechanism of action and resistance pathways. *J Antimicrob Chemother* 57: 619–627

59 Zollner B, Feucht HH, Schroter M, Schafer P, Plettenberg A, Stoehr A, Laufs R (2001) Primary genotypic resistance of HIV-1 to the fusion inhibitor T-20 in long-term infected patients. *AIDS* 15: 935–936

60 Holguin A, Faudon JL, Labernardiere J-L, Soriano V (2006) Susceptibility of HIV-1 non-B subtypes and recombinant variants to Enfuvirtide. *J Clin Virol* 38: 176–180

61 Stanfield-Oakley S, Jeffrey J, McDanal C, Mosier S, Talton L, Jin L, Sista P, Cammack N, Matthews TJ, Greenberg ML (2003) Determinants of susceptibility to enfuvirtide map to gp41 in enfuvirtide-naïve HIV-1. *Antiviral Ther* 8: S22

62 Heil ML, Decker JM, Sfakianos JN, Shaw GM, Hunter E, Derdeyn CA (2004) Determinants of human immunodeficiency virus type 1 baseline susceptibility to the fusion inhibitors enfuvirtide and T-649 reside outside the peptide interaction site. *J Virol* 78: 7582–7589

63 Derdeyn CA, Decker JM, Sfakianos JN, Wu X, O'Brien WA, Ratner L, Kappes JC, Shaw GM, Hunter E (2000) Sensitivity of human immunodeficiency virus type 1 to the fusion inhibitor T-20 is modulated by coreceptor specificity defined by the V3 loop of gp120. *J Virol* 74: 8358–8367

64 Derdeyn CA, Decker JM, Sfakianos JN, Zhang Z, O'Brien WA, Ratner L, Shaw GM, Hunter E (2001) Sensitivity of human immunodeficiency virus type 1 to fusion inhibitors targeted to the gp41 first heptad repeat involves distinct regions of gp41 and is consistently modulated by gp120 interactions with the coreceptor. *J Virol* 75: 8605–8614

65 Reeves JD, Gallo SA, Ahmad N, Miamidian JL, Harvey PE, Sharron M, Pohlmann Sfakianos JN, Derdeyn CA, Blumenthal R, Hunter E, Doms RW (2002) Sensitivity of HIV-1 to entry inhibitors correlates with envelope/coreceptor affinity, receptor density, and fusion kinetics. *Proc Natl Acad Sci USA* 99: 16249–16254

66 Melby T, DeSpirito M, DeMasi R, Heilek-Snyder G, Greenberg ML, Graham N (2006) HIV-1 coreceptor use in triple-class treatment-experienced patients: baseline prevalence, correlates, and relationship to enfuvirtide response. *J Infect Dis* 194: 238–246

67 Lu J, Sista P, Giguel F, Greenberg M, Kuritzkes DR. (2004) Relative replicative fitness of human immunodeficiency virus type 1 mutants resistant to enfuvirtide (T-20). *J Virol* 78: 4628–4637

68 Menzo S, Castagna A, Monachetti A, Hasson H, Danise A, Carini E, Bagnarelli P, Lazzarin A, Clementi M (2004) Genotype and phenotype patterns of human immunodeficiency virus type 1 resistance to enfuvirtide during long-term treatment. *Antimicrob Agents Chemother* 48: 3253–3259

69 Deeks SG, Lu J, Hoh R, Neilands TB, Beatty G, Huang W, Leigler T, Hunt P, Martin JN, Kuritzkes DR (2007) Interruption of enfuvirtide in HIV-1-infected adults with incomplete viral suppression on an enfuvirtide-based regimen. *J Infect Dis* 195: 387–391

70 Xu L, Pozniak A, Wildfire A, Stanfield-Oakley S, Mosier SM, Ratcliffe D, Workman J, Joall A, Myers R, Smit E et al. (2005) Emergence and evolution of enfuvirtide resistance following long-term therapy involves heptad repeat 2 mutations within gp41. *Antimicrob Agents Chemother* 49: 1113–1119

71 Poveda E, Rodes B, Labernardiere J-L, Benito JM, Toro C, Gonzalez-Lahoz J, Faudon J-L, Clavel F, Schapiro J, Soriano V (2004) Evolutions of genotypic and phenotypic resistance to enfuvirtide

in HIV-infected patients experiencing prolonged virologic failure. *J Med Virol* 74: 21–28

72 Mink M, Mosier SM, Janumpalli S, Davison D, Jin L, Melby T, Sista P, Erickson J, Lambert D, Stanfield-Oakley SA et al. (2005) Impact of human immunodeficiency virus type 1 gp41 amino acid substitutions selected during enfuvirtide treatment on gp41 binding and antiviral potency of enfuvirtide *in vitro*. *J Virol* 79: 12447–12454

73 Bai X, Wilson KL, Seedorff JE, Ahrens D, Green J, Davison DK, Lin J, Stanfield-Oakley SA, Mosier S, Melby T et al. (2006) The impact of the N43D resistance mutation on enfuvirtide sensitivity and six-helix bundle structure in combination with the E137K polymorphism. *Antiviral Ther* 11: S55

74 Reeves JD, Lee F-H, Miamidian JL, Jabara CB, Juntilla MM, Doms RW (2005) Enfuvirtide resistance mutations: impact on human immunodeficiency virus envelope function, entry inhibitor sensitivity, and virus neutralization. *J Virol* 79: 4991–4999

75 Baldwin CE, Sanders RW, Den Y, Jurriaans S, Lange JM, Lu M, Berkhout B (2004) Emergence of drug-dependent human immunodeficiency virus type 1 variant during therapy with the T20 fusion inhibitor. *J Virol* 78: 12428–12437

76 Etemad-Moghadam B, Rhone D, Steenbeke T, Sun Y, Manola J, Gelman R, Fanton JW, Racz P, Tenner-Racz K, Axthelm MK et al. (2001) Membrane-fusing capacity of the human immunodeficiency virus envelope proteins determines the efficiency of CD^+ T-cell depletion in macaques infected by a simian-human immunodeficiency virus. *J Virol* 75: 5646–5655

77 Aquaro S, D'Arrigo R, Svicher V, Perri GD, Caputo SL, Visco-Comandini U, Santoro M, Bertoli A, Mazzotta F, Bonora S et al. (2006) Specific mutations in HIV-1 gp41 are associated with immunological success in HIV-1-infected patients receiving enfuvirtide treatment. *J Antimicrob Chemother* 58: 714–722

78 Melby T, DeSpirito M, DeMasi R, Heilik-Snyder G, Thommes J, Greenberg ML, Graham N (2006) Impact of enfuvirtide resistance genotype on CD4 increases in patients with ongoing viral replication while receiving enfuvirtide. *Antiviral Ther* 11: S43

79 Barretina J, Blanco J, Bonjoch A, Llano A, Clotet B, Este JA (2004) Immunological and virological study of enfuvirtide-treated HIV-positive patients. *AIDS* 18:1673–1682

80 Meissner EG, Zhang L, Jiang S, Su L (2006) Fusion-induced apoptosis contributes to thymocyte depletion by a pathogenic human immunodeficiency virus type 1 envelope in the human thymus. *J Virol* 80: 11019–11030

81 Garg H, Blumenthal R (2005) HIV gp41-induced apoptosis is mediated by caspase-3-dependent mitochondrial depolarization, which is inhibited by HIV protease inhibitor nelfinavir. *J Leukoc Biol* 79: 352–362

82 Grinsztejn B, Nguyen B, Katlama C, Gatell J, Lazzarin A, Vittecoq D, Gonzales C, Chen J, Isaacs R, and the Protocol 1005 Study Team (2006) Potent efficacy of MK-0518, a novel HIV-1 integrase inhibitor, in patients with triple-class resistant virus: 24-week data. Program and abstracts of the 46th Interscience Conference on Antimicrobial Agents and Chemotherapy; September 27–30, 2006; San Francisco, CA, abstr. no. H-1670b

83 Zhang X-Q, Sorensen M, Fung M, Schooley RT (2006) Synergistic *in vitro* antiretroviral activity of a humanized monoclonal anti-CD4 antibody (TNX-355) and enfuvirtide (T-20). *Antimicrob Agents Chemother* 50: 2231–2233

Entry Inhibitors in HIV Therapy
Edited by Jacqueline D. Reeves and Cynthia A. Derdeyn

Targets for drug development – past and present

Roy M. Gulick

Weill Medical College of Cornell University, Cornell HIV Clinical Trials Unit, Box 566, 525 East 68th Street, New York 10021, USA

Targets for HIV drug development – Traditional classes (Tab. 1)

The first cases of AIDS were recognized in 1981 [1]. With the realization that patients with AIDS experienced nearly universal mortality, the search for effective treatment began urgently. This search was facilitated greatly by the discovery of the causative agent, the human immunodeficiency virus (HIV) [2, 3]. It was recognized early on that the replication cycles of other animal and human retroviruses depended on the virus-specific enzyme reverse transcriptase, and this became the first HIV drug target. Several compounds with demonstrated activity against other retroviruses were reported to have activity against HIV *in vitro* by inhibiting the reverse transcriptase enzyme, such as suramin [4] and ribavirin [5]. However, clinical trials of these agents ultimately showed no clinical benefits in HIV-infected patients [6, 7].

The synthesis of the thymidine analogue, later named zidovudine, was reported originally by investigators at the Detroit Institute of Cancer Research as part of a drug discovery program for new cancer chemotherapeutic agents [8]. Lin and Prusoff [9], investigators at Yale, first reported the *in vitro* activity of thymidine analogues as antiviral agents. Mitsuya and colleagues [10] from the National Cancer Institute and Wellcome Research Laboratories reported the first data that zidovudine triphosphate inhibited HIV replication *in vitro* and advocated its "cautious exploration" in HIV-infected patients.

Shortly thereafter, zidovudine entered clinical trials in patients with symptomatic HIV disease or AIDS and proved to confer a significant short-term survival benefit: in the first clinical trial sponsored by Burroughs Wellcome and led by academic investigators from the University of Miami, it was found that 1 of 145 subjects who received zidovudine, compared to 19 of 137 subjects who received placebo, died over 8–24 weeks ($p < 0.001$) [11]. Together with other studies, this led to the approval of zidovudine as the first therapy for HIV infection in 1987. Zidovudine continues to be used today as a component of HIV treatment regimens. Additional nucleoside analogues were developed subsequently, and currently, eight nucleoside (or nucleotide) analogues are approved for the treatment of HIV infection.

Table 1. Approved antiretroviral drugs: 2007

Class	Year of FDA approval
HIV reverse transcriptase inhibitors:	
Nucleosides/nucleotides (NRTI):	
zidovudine	1987
didanosine	1991
zalcitabine*	1992
stavudine	1994
lamviudine	1995
abacavir	1998
tenofovir	2001
emtricitabine	2003
Non-nucleosides (NNRTI):	
nevirapine	1996
delavirdine	1997
efavirenz	2000
HIV protease inhibitors (PI):	
saquinavir	1995
ritonavir	1996
indinavir	1996
nelfinavir	1997
amprenavir**	1999
lopinavir/ritonavir	2000
fosamprenavir	2003
atazanavir	2003
tipranavir	2005
darunavir	2006
HIV entry inhibitors (EI):	
enfuvirtide***	2003

* no longer available; ** available only as a liquid; *** requires subcutaneous injection.

A second class of reverse transcriptase inhibitors, the non-nucleosides (NNRTI), was first explored in the late 1980s. Scientists at Boehringer-Ingelheim reported a series of dipyridodizepinone compounds, including nevirapine, that demonstrated potent activity *in vitro* against HIV reverse transcriptase [12] but bound to the enzyme at a site distinct from the nucleoside analogues [13]. This distinct NNRTI binding site enabled nevirapine to have *in vitro* antiretroviral activity against viral strains with resistance to nucleoside analogues such as zidovudine, and, in addition, likely explained the *in vitro* antiretroviral synergy between the two drug classes [14]. Although, in NIH-sponsored AIDS Clinical Trials Group (ACTG) trials, resistance to nevirapine developed quickly when the drug was given alone [15], combination therapy with nevirapine and two nucleoside analogues ultimately demonstrated potent, durable antiretroviral activity in industry-sponsored studies [16], leading to the approval of nevirapine in 1996. Currently, three NNRTI are approved for the treatment of HIV infection and NNRTI-based combination regimens are a mainstay of HIV therapy.

The second mechanistically distinct class of HIV drugs are the protease inhibitors. Investigators at Kyushu University School of Medicine and Kitashita University reported that the N-terminal end of the polymerase gene of retroviruses coded for an aspartyl protease [17]. Subsequently, Kramer and colleagues from Roche and the National Cancer Institute reported that the HIV gag protein was processed by a protease, and suggested this enzyme as a target for HIV drug development [18]. The three-dimensional structure of the HIV protease enzyme was described by scientists at Merck, who identified its homodimer structure and active site and also suggested it as a target for drug therapy [19].

Groups at the National Institute of Allergy and Infectious Diseases and Upjohn identified candidate HIV protease inhibitors that demonstrated antiretroviral activity by inhibiting proteolysis of the HIV-1 gag polyprotein p55 to the structural proteins p24 and p17 [20, 21]. Scientists at Roche described a series of peptide derivatives that mimicked the transition-state of the HIV polyproteins and potently inhibited the protease enzyme [22]. Subsequently, saquinavir, ritonavir, and indinavir were identified as compounds that potently inhibited the HIV protease *in vitro* [23–25]. Clinical trials sponsored both by the National Institutes of Health (NIH) and industry, and conducted by academic clinical researchers, showed these compounds had potent antiretroviral activity [26–28] and conferred clinical benefits, including reductions in HIV-related morbidity and mortality [29, 30], particularly when used in combination with nucleoside analogues. These results led to the approval in 1995–1996 of saquinavir, ritonavir, and indinavir. The widespread use of protease inhibitor-based combination therapy followed, leading in turn to dramatic reductions in HIV-related morbidity and mortality in developed countries around the world [31, 32]. Currently there are ten HIV protease inhibitors approved for the treatment of HIV infection.

Despite the availability of nucleoside analogues, NNRTIs, and protease inhibitors, some patients ultimately develop multidrug-resistant virus: for example, Richman and colleagues [33] reported that an estimated 63% of the over 130 000 Americans who received care in 1996 had HIV RNA levels greater than 500 copies/ml by 1998 and that 76% of those patients with detectable viremia had resistance to one or more antiretroviral drugs. With this in mind, the search for compounds with new mechanisms of action continued and intensified.

Targets for HIV drug development – HIV entry

Attachment inhibitors

Soon after the discovery of HIV, it was appreciated that the $CD4^+$ T lymphocyte was the target cell of the virus and that binding to the CD4 receptor was required for viral entry into the cell [34, 35]. Thus, HIV entry became an early potential target for HIV drug development. Recognition that HIV bound to the CD4 receptor through the viral envelope glycoprotein gp120 [36] led to the identification of both the CD4 receptor and gp120 as the first specific targets of HIV entry inhibition. In the late 1980s, a number of groups from industry and academia reported that recombinant, soluble CD4 (rsCD4) was a potent inhibitor of HIV replication *in vitro* [37–41], and suggested its use as therapy for HIV infection.

Subsequently, academic investigators reported a Phase I/II clinical trial of rsCD4 sponsored by the NIH-funded ACTG [42]. In this 28-day study of rsCD4 given intravenously or intramuscularly, they reported the compound had a short half-life and only modest antiretroviral effect at the highest doses tested. Other studies later confirmed that only very high doses of rsCD4 produced an antiretroviral effect [43–45], and that this likely was due to decreased affinity of rsCD4 for clinical (*versus* laboratory) viral isolates [46]; as a result, this therapeutic approach was set aside.

Chemokine receptor inhibitors: CCR5

In the early 1990s, investigators from the Amsterdam Cohort Study reported that clinical HIV strains that induced syncytia formation *in vitro* (SI) in MT-2 T cells were associated with more rapid CD4 cell declines and increased clinical progression, compared to non-syncytia-inducing (NSI) viral isolates, and that a change from NSI to SI viral phenotype was associated with both accelerated CD4 decline and clinical progression [47]. Investigators at the National Cancer Institute reported the antiretroviral effects of chemokines (RANTES, MIP-1 alpha, and MIP-1 beta) and first suggested that using chemokines or their analogues could be a therapeutic strategy for HIV infection [48]. Shortly thereafter, investigators from five different groups reported that the chemokine receptors CXCR4 (bound by SI, or X4-virus) and CCR5 (bound by NSI, or R5-virus) were required for HIV binding and entry into the CD4 cell [49–53]. This paved the way for the identification and development of specific chemokine receptor inhibitors directed against either the CCR5 or CXCR4 receptors on CD4 cells.

Among several compounds identified that inhibited CCR5-mediated HIV entry *in vitro*, SCH 351125 (Schering C) took the lead as an orally bioavailable, small-molecule CCR5 inhibitor that demonstrated potent antiretroviral activity both *in vitro* and in a mouse model [54]. The first human trial of a chemokine receptor inhibitor was a 14-day study of SCH 351125 sponsored by Schering-Plough and conducted by investigators in the U.S. and France [55]. Although the compound demonstrated antiretroviral activity, further clinical development was abandoned due to dose-dependent cardiac QT interval prolongation [56].

Chemokine receptor inhibitors: CXCR4

In the early 1990s, investigators at the Rega Institute in Belgium first described the potent *in vitro* antiretroviral activity of the bicyclams, a class of compounds that inhibited an early event in the life cycle of HIV [57]. This same group identified a specific bicyclam (first called JM-3100 and later AMD-3100) that demonstrated potent antiretroviral activity *in vitro* and in animal studies, as well as inhibition of viral-induced syncytium formation [58]. Several groups subsequently reported that AMD-3100 specifically interacted with the CXCR4 co-receptor [59, 60]. An early clinical study sponsored by Anormed and conducted by investigators at Johns Hopkins assessed the safety and pharmacokinetics of parenteral and oral AMD-3100 in HIV-negative volunteers but found no detectable drug following oral dosing [61]. In a subsequent study, AMD-3100 was administered intravenously for 14 days to 40 HIV-infected patients, 30% of whom had virus with SI phenotype using the MT-2 cell assay [62]. In this study, two patients experienced premature ventricular contractions, most patients at the higher doses experienced paresthesias, and only 1 patient with CXCR4-using virus, who also received the highest AMD-3100 dose, demonstrated a significant decrease in HIV RNA level of 0.9 log copies/ml. Concerns about suboptimal antiretroviral potency led to discontinuation of development of the compound as an antiretroviral agent.

Fusion inhibitors

In 1989, scientists at Louisiana State University first proposed a helical model of the gp41 transmembrane protein of HIV, based on the known structure of the HA2 transmembrane protein of influenza [63]. Specific helical regions of gp41 known as HR (heptad repeat) 1 and 2 were described [63, 64]. These regions were deduced to interact with one another in a coiled-coil reaction by investigators at Duke, who identified peptides that bound to these sequences and inhibited viral and cellular membrane fusion [65, 66]. These peptides demonstrated potent antiretroviral activity *in vitro* and led to the development by scientists at Duke and Trimeris of a related 36 amino acid peptide, enfuvirtide (for <u>en</u>velope <u>fu</u>sion <u>vir</u>al pep<u>tide</u>), derived from the HR 2 sequence of gp41 that bound to the HR 1 sequence and prevented membrane fusion *in vitro*. Subsequently, enfuvirtide was proposed for clinical testing as the first HIV fusion inhibitor [67].

In an early clinical study sponsored by Trimeris, Inc. and led by investigators at the University of Alabama, intravenous enfuvirtide at higher doses proved to have potent antiretroviral activity in HIV-infected patients over 14 days [68]. Subsequently, Phase III studies sponsored by Trimeris and Roche and conducted by academic investigators in North America, Europe, and Australia demonstrated potent antiretroviral activity of twice-daily subcutaneous enfuvirtide (along with an optimized antiretroviral regimen) in treatment-experienced patients [69, 70]. On the basis of these studies, enfuvirtide was approved in 2003 as the first HIV fusion inhibitor.

HIV entry inhibitors: Current status and challenges (Tab. 2)

Today, three-drug combination antiretroviral therapy is the standard of care for the treatment of HIV infection [71, 72]. Among the 22 approved antiretroviral drugs, the only entry inhibitor approved to date is enfuvirtide. Although potent, enfuvirtide is a parenteral agent requiring twice-daily subcutaneous dosing and was tested in heavily treatment-experienced patients. As a result, it is specifically labeled "for the treatment of HIV-1 infection in treatment-experienced patients with evidence of HIV-1 replication despite ongoing antiretroviral therapy" [73]. While a number of HIV entry inhibitors are currently under investigation, these compounds face challenges both in their clinical development and in finding a place in the current HIV treatment paradigm.

Table 2. HIV entry inhibitors and stage of development (partial list)

Stage of development	CD4 attachment inhibitors	Chemokine receptor inhibitors	Fusion inhibitors
Approved	–	–	enfuvirtide
Phase II/III	–	maraviroc (R5)	–
Phase II	TNX-355	vicriviroc (R5)	–
Phase I/II	PRO 542	–	–
Phase I	–	AMD-070 (X4) INCB9471 (R5) CCR5mAb004 (R5) PRO 140 (R5) TAK-652 (R5)	sifuvirtide
Preclinical	BMS small-molecule inhibitors	AMD-887 (R5) KRH-3140 (X4) KRH-3955 (X4)	TRI-291144
No longer in development	rsCD4 BMS-378806 BMS-488403	AMD-3100 (X4) aplaviroc (R5) SCH 351125 (R5) TAK-779 (R5)	T-1249 TRI-290999

Attachment inhibitors

Several attachment inhibitors are currently in clinical development: for example, PRO 542 (CD4-IgG2) is an antibody-like fusion protein developed by scientists at Progenics Pharmaceuticals that binds to gp120 and blocks attachment to the CD4 receptor [74]. The compound demonstrated antiretroviral activity with single doses up to 25 mg/kg in adults in studies led by investigators at Mt. Sinai [75, 76], and four weekly doses of 10 mg/kg in children in a Pediatric ACTG study [77]. However, it has been challenging to accrue patients to Phase II studies requiring three times weekly parenteral dosing and, as a result, development of PRO 542 has been slow.

TNX-355 is a humanized IgG4 monoclonal antibody developed by scientists at Biogen that binds to the second extracellular domain of the CD4 receptor [78]. The compound demonstrated dose-related antiretroviral activity in a Phase I study of HIV-infected subjects receiving single doses [79], and in a Phase Ib study of HIV-infected subjects receiving various doses over 9 weeks [80]. More recently, Tanox reported 48-week results from a Phase II study in triple-class treatment-experienced patients that showed that the two doses of TNX-355 tested (each given with optimized background antiretrovirals), had significantly better antiretroviral activity than placebo and were well tolerated [81]. As a monoclonal antibody, TNX-355 must be given parenterally, but its long half-life allows once- or twice-weekly dosing. Follow-up in a larger number of patients will be required to further assess both the longer-term virologic activity of the compound and any immunologic consequences of binding the CD4 receptor.

The first representative of a series of small-molecule compounds that bind within the CD4-binding pocket of gp120 and prevent gp120-CD4 interaction, BMS-378806, was synthesized and reported by scientists at Bristol-Myers Squibb [82, 83]. This compound showed potent antiretroviral activity *in vitro*, but did not achieve target exposures in HIV-uninfected adults. A follow-on compound, BMS-488043, demonstrated adequate pharmacokinetics in HIV-uninfected volunteers [84] and antiretroviral activity in a clinical study over 8 days of dosing [85]. Although further development of BMS-488043 was halted, additional small-molecule compounds are in development.

Chemokine receptor inhibitors: CCR5

Three small-molecule CCR5 inhibitors, all noncompetitive allosteric antagonists of the CCR5 receptor, have reached advanced clinical development: aplaviroc, maraviroc, and vicriviroc. Researchers at GlaxoSmithKline reported that aplaviroc was a potent antagonist of the CCR5 receptor [86] with potent antiretroviral activity *in vitro* [87]. Short-term safety and dose-dependent pharmacokinetics were demonstrated in HIV-uninfected volunteers [88]. A Phase I study of aplaviroc given for 10 days in 40 HIV-infected subjects showed dose-dependent antiretroviral activity and no safety concerns [89]. However, upon initiation of larger Phase II studies in HIV-infected treatment-naïve and treatment-experienced patients, severe hepatoxicity attributed to the compound developed in at least 5 patients, and further development of aplaviroc was stopped in October 2005 [90].

Scientists at Pfizer screened their compound library to identify small-molecule CCR5 antagonists, found a candidate imidazopyridine compound (UK-107,543), and then synthesized and profiled nearly 1000 analogues to optimize receptor binding potency, antiretroviral activity, and pharmacokinetic properties; through this process, maraviroc was identified as the lead clinical candidate [91]. Compound screening was further enhanced by the use of a high-throughput binding assay for the hERG potassium channel, to reduce the chances of QT interval prolongation [92]. Studies in HIV-uninfected volunteers were conducted to explore pharmacokinetics and drug-drug interactions [93]. Investigators conducting the first study in HIV-infected patients gave maraviroc at eight different dosing schemes to 63 participants and reported dose-dependent antiretroviral activity and no safety concerns [94]. With further clinical development, two cases of hepatotoxicity in patients taking maraviroc were reported, but neither was clearly related to the study drug [95].

Phase II/III studies of maraviroc in both treatment-naïve and -experienced patients with R5-virus are now fully enrolled and results are anticipated. However, in early 2006, the Data Safety Monitoring Board recommended stopping the lower dose of maraviroc (300 mg once-daily) in the treatment-naïve study due to suboptimal antiretroviral activity compared to efavirenz (each given with two nucleosides) [96].

Scientists at Schering, building on their experiences with SCH 351125 (see above), tested additional compounds for antiretroviral, pharmacokinetic, and reduced hERG channel-blocking properties; they identified vicriviroc (Schering D) as the lead clinical candidate [56]. Vicriviroc is structurally distinct from SCH 351125 but also demonstrated potent antiretroviral activity *in vitro*. The first study in HIV-infected patients (not taking other antiretrovirals), demonstrated dose-dependent activity with vicriviroc administered at 10, 25, and 50 mg over 14 days without safety concerns [97]. Pharmacokinetic studies in HIV-uninfected patients demonstrated that ritonavir enhanced the concentrations of vicriviroc by 3.5–5-fold [98]. These data supported Phase II studies in both treatment-naïve (without concomitant ritonavir) and treatment-experienced (with ritonavir) patients.

Late in 2005, the Data Safety Monitoring Board recommended stopping the vicriviroc treatment-naïve study due to suboptimal antiretroviral activity compared to efavirenz (each in combination with two nucleosides) [99]. In the ACTG 5211 study of treatment-experienced patients, vicriviroc at 10 and 15 mg doses (together with a ritonavir-based regimen) demonstrated durable virologic activity over 24 weeks [100]. However, four lymphomas and one gastric adenocarcinoma occurred among the 98 patients taking vicriviroc on this study, although causality of vicriviroc was uncertain. Additional Phase II studies of vicriviroc are underway.

A 32-base pair deletion in the gene that codes for the CCR5 receptor is found in 2–5% of populations in Europe, the Middle East, and the Indian subcontinent [101], and such individuals appear to have normal immune function and high-level resistance to HIV infection [102]. Subsequent studies suggested associations between the CCR5 deletion and clinical implications, such as reduced signs and symptoms of rheumatoid arthritis [103], less inflammation and fibrosis from HCV [104], and longer graft survival following renal transplant [105]. In contrast, there are reported associations between the CCR5 deletion and decreased graft survival following liver transplant [106], and, in the setting of West Nile virus infection, increased severity of disease and death [107]. Congenital absence of CCR5 likely is not the same as pharmacologic blocking of the receptor, and there may be long-term immunologic consequences associated with the use of this class of compounds.

A theoretical risk in using CCR5 inhibitors is selection for X4-virus. Isolated cases have been described in the studies of patients with R5-virus with aplaviroc [108] and vicriviroc [97], but the best documented cases were in the study of maraviroc: in this study, two patients with R5-virus at baseline developed detectable X4-virus at day 11 that appeared to be due to outgrowth of a minor population of pre-existing X4-virus, and then reverted to R5-virus only following discontinuation of maraviroc [109]. In addition, the effect of a CCR5 inhibitor in 186 patients with dual-tropic or mixed viral populations (both R5- and X4-virus) was assessed in an initial 24-week study of maraviroc (*versus* placebo); no clear virologic effect of the CCR5 inhibitor was found, but increases in CD4 cell count were demonstrated [110].

Chemokine receptor inhibitors: CXCR4

Following discontinuation of the clinical development of parenteral AMD-3100, the same team of investigators identified an orally bioavailable CXCR4 inhibitor, AMD-070, with potent antiretroviral activity against X4-viruses *in vitro* [111]. The first clinical study of AMD 070, ACTG 5191, reported HIV-uninfected volunteers who received single oral doses of the compound; AMD 070 was found to have dose-proportional pharmacokinetics and was generally well tolerated [112]. Studies in HIV-infected patients are in progress; HIV RNA reductions of 1.3 log copies/ml in four of eight patients receiving the lowest tested AMD 070 dose twice-daily for 10 days were reported in preliminary results [113].

Fusion inhibitors

While effective, enfuvirtide must be administered by subcutaneous injection twice daily. In the presence of ongoing viremia, enfuvirtide selected for drug-resistant strains [114]. Thus, the search for additional fusion inhibitors continues. Researchers at Trimeris initially identified a second peptide that bound to gp41 and prevented fusion, T-1249, which could be administered once-daily, and demonstrated dose-dependent antiretroviral activity both in HIV-infected patients naïve to enfuvirtide [115] and patients who experienced virologic failure on enfuvirtide [108]. However, clinical development was suspended due to formulation issues. Sifuvirtide is a peptide fusion inhibitor, developed by scientists in China at FusoGen Pharmaceuticals, that demonstrated a longer half-life than enfuvirtide in a study in monkeys [116], and recently entered clinical studies. Also, researchers at Trimeris reported a new peptide, TRI-291144, that is undergoing evaluation with a sustained-release formulation targeting once-weekly dosing [117]. This peptide demonstrated potent antiretroviral activity *in vitro*, including against enfuvirtide-resistant viral strains, and in monkeys, and clinical studies are anticipated.

HIV entry inhibitors: What's next?

Proven, durable antiretroviral efficacy and drug safety are key considerations for selecting antiretroviral regimens today from among the approved HIV drugs. Although our current antiretroviral therapies are effective in many patients and generally safe and well tolerated, there is room for improvement. Newer agents that are more convenient, less toxic, active against drug-resistant viral strains, or less expensive could have an important role to play in HIV treatment. Both general and class-specific factors impact how HIV entry inhibitors may fit into the HIV treatment paradigm in the future.

The risk-benefit ratios for choosing individual agents vary for HIV-infected treatment-naïve and treatment-experienced patients. Current treatment guidelines recommend initial HIV treatment with two nucleosides and either an NNRTI or a boosted protease inhibitor [71, 72], based on published data with these regimens demonstrating potent antiretroviral efficacy and proven safety of the drug regimens. With years of available efficacy and safety data, it will be challenging for any new agent to replace the current proven initial antiretroviral drugs. The two studies of treatment-naïve patients that compared a standard of care regimen to a maraviroc- or vicriviroc-based regimen and found suboptimal potency of the CCR5 inhibitors at lower doses [96, 99] exemplify this issue. Also, the lack of safety data in large numbers of patients or for extended periods of time with the newer investigational compounds raises issues. Reports of drug toxicity with chemokine receptor inhibitors (including QT interval prolongation, hepatotoxicity and concerns about immunologic consequences including malignancies) illustrate this point. Thus, long-term follow-up of patients taking HIV entry inhibitors will be required to assess toxicities carefully over time.

For treatment-experienced patients with few, if any, options for effective antiretroviral therapy, the risk-benefit ratio of taking a new agent differs. Therefore, it is appropriate that many of the HIV entry inhibitors have initiated clinical development in this population. When a suitable amount of antiretroviral activity and safety information is available, it is then reasonable to pursue testing in patients at earlier stages of HIV infection. From a pathogenesis point of view, R5-virus is much more common in acute, early, and chronic infection prior to treatment, which supports testing of CCR5 inhibitors in these clinical settings. Given their mechanism of action, HIV entry inhibitors also could be useful (provided safety is established) as post-exposure prophylaxis or even pre-exposure prophylaxis in HIV-uninfected individuals, as suggested by recent animal data [118].

Parenteral dosing of antiretroviral agents also poses challenges to the HIV entry inhibitor class, specifically for the monoclonal antibodies that inhibit CD4 receptor attachment and the peptides that inhibit membrane fusion. Based on the experience with enfuvirtide, compounds requiring parenteral dosing once- or twice-daily will be reserved exclusively for patients with extensive treatment experience and multidrug-resistant virus, rather than patients at earlier stages of HIV infection, purely for practical reasons. Even for the advanced group of patients, other investigational compounds with new mechanisms of action in development that may be given orally, such as the HIV integrase inhibitors [119, 120] and HIV maturation inhibitors [121] likely will be preferred, because they will be easier to use than a parenteral agent. Whether any parenteral therapy, even one dosed infrequently, will gain widespread acceptance in the HIV treating community is not certain.

Combinations of the different classes of HIV entry inhibitors demonstrate synergy *in vitro* [122, 123], which supports clinical testing of these combinations. In addition, based on the mechanism of action, it is reasonable to speculate that patients who have enfuvirtide-resistant viral strains could experience enhanced susceptibility when a CCR5 inhibitor is co-administered [124]. Also, the concomitant use of a CCR5 inhibitor and a CXCR4 inhibitor demonstrated activity *in vitro* against R5-, X4-, and dual-mixed viral strains [125] and could be explored clinically. Combination studies of HIV entry inhibitors both for entry inhibitor-naïve and enfuvirtide-experienced patients are in progress.

Some class-specific issues also pose special challenges to the entrance of HIV entry inhibitors into the current HIV treatment paradigm. To date, CD4 attachment inhibitors and fusion inhibitors have demonstrated bioavailability and formulation challenges; chemokine receptors may well require a special assay to determine viral tropism prior to treatment [126]; CCR5 inhibitors may be associated with co-receptor change (e.g., to X4-virus) and immunologic consequences; the one approved fusion inhibitor, enfuvirtide, appears to select readily for drug resistance in the setting of ongoing viremia [127]; and all of the HIV entry inhibitor class compounds have been associated with a variety of short-term drug-specific toxicities. Nevertheless, with a novel mechanism of action, HIV entry inhibitors could provide a valuable addition to the HIV treatment armamentarium.

Acknowledgements
Thanks to John Moore, PhD, for thoughtful review and comments.

References

1 Centers for Disease Control (1981) Pneumocystis pneumonia – Los Angeles. *MMWR* 30: 250–252
2 Barre-Sinoussi F, Chermann JC, Rey F, Nugeyre MT, Chamaret S, Gruest J, Dauguet C, Axler-

Blin C, Vezinet-Brun F, Rouzioux C et al. (1983) Isolation of a T-lymphotropic retrovirus from a patient at risk for acquired immune deficiency syndrome (AIDS). *Science* 220: 868–871

3 Gallo RC, Sarin PS, Gelmann EP, Robert-Guroff M, Richardson E, Kalyanaraman VS, Mann D, Sidhu GD, Stahl RE, Zolla-Pazner S et al. (1983) Isolation of human T-cell leukemia virus in acquired immune deficiency syndrome (AIDS). *Science* 220: 865–867

4 Mitsuya H, Popovic M, Yarchoan R, Matsushita S, Gallo RC, Broder S (1984) Suramin protection of T cells *in vitro* against infectivity and cytopathic effect of HTLV-III. *Science* 226: 172–174

5 McCormick JB, Getchell JP, Mitchell SW, Hicks DR (1984) Ribavirin suppresses replication of lymphadenopathy-associated virus in cultures of human adult T lymphocytes. *Lancet* 2: 1367–1369

6 Kaplan LD, Wolfe PR, Volberding PA, Feorino P, Levy JA, Abrams DI, Kiprov D, Wong R, Kaufman L, Gottlieb MS (1987) Lack of response to suramin in patients with AIDS and AIDS-related complex. *Am J Med* 82: 615–620

7 Spector SA, Kennedy C, McCutchan JA, Bozzette SA, Straube RG, Connor JD, Richman DD (1989) The antiviral effect of zidovudine and ribavirin in clinical trials and the use of p24 antigen levels as a virologic marker. *J Infect Dis* 159: 822–828

8 Horwitz JP, Chua J, Noel M (1964) Nucleosides V – the monomesylates of 1-2'-deoxy-B-D-lyxofuranosyl thymine. *J Org Chem* 29: 2076–2078

9 Lin TS, Prusoff WH (1978) Synthesis and biological activity of several amino analogues of thymidine. *J Med Chem* 21: 109–112

10 Mitsuya H, Weinhold KJ, Furman PA, St Clair MH, Lehrman SN, Gallo RC, Bolognesi D, Barry DW, Broder S (1985) 3'-Azido-3'-deoxythymidine (BW A509U): an antiviral agent that inhibits the infectivity and cytopathic effect of human T-lymphotropic virus type III/lymphadenopathy-associated virus *in vitro*. *Proc Natl Acad Sci USA* 82: 7096–7100

11 Fischl MA, Richman DD, Grieco MH, Gottlieb MS, Volberding PA, Laskin OL, Leedom JM, Groopman JE, Mildvan D, Schooley RT et al. (1987) The efficacy of azidothymidine (AZT) in the treatment of patients with AIDS and AIDS-related complex. A double-blind, placebo-controlled trial. *N Engl J Med* 317: 185–191

12 Merluzzi VJ, Hargrave KD, Labadia M, Grozinger K, Skoog M, Wu JC, Shih CK, Eckner K, Hattox S, Adams J et al. (1990) Inhibition of HIV-1 replication by a nonnucleoside reverse transcriptase inhibitor. *Science* 250: 1411–1413

13 Wu JC, Warren TC, Adams J, Proudfoot J, Skiles J, Raghavan P, Perry C, Potocki I, Farina PR, Grob PM (1991) A novel dipyridodiazepinone inhibitor of HIV-1 reverse transcriptase acts through a nonsubstrate binding site. *Biochemistry* 30: 2022–2026

14 Richman D, Rosenthal AS, Skoog M, Eckner RJ, Chou TC, Sabo JP, Merluzzi VJ (1991) BI-RG-587 is active against zidovudine-resistant human immunodeficiency virus type 1 and synergistic with zidovudine. *Antimicrob Agents Chemother* 35: 305–308

15 Richman DD, Havlir D, Corbeil J, Looney D, Ignacio C, Spector SA, Sullivan J, Cheeseman S, Barringer K, Pauletti D et al. (1994) Nevirapine resistance mutations of human immunodeficiency virus type 1 selected during therapy. *J Virol* 68: 1660–1666

16 Montaner JS, Reiss P, Cooper D, Vella S, Harris M, Conway B, Wainberg MA, Smith D, Robinson P, Hall D et al. (1998) A randomized, double-blind trial comparing combinations of nevirapine, didanosine, and zidovudine for HIV-infected patients: the INCAS Trial. *JAMA* 279: 930–937

17 Toh H, Kikuno R, Hayashida H, Miyata T, Kugimiya W, Inouye S, Yuki S, Saigo K (1985) Close structural resemblance between putative polymerase of a Drosophila transposable genetic element 17.6 and pol gene product of Moloney murine leukaemia virus. *EMBO J* 4: 1267–1272

18 Kramer RA, Schaber MD, Skalka AM, Ganguly K, Wong-Staal F, Reddy EP (1986) HTLV-III gag protein is processed in yeast cells by the virus pol-protease. *Science* 231: 1580–1584

19 Navia MA, Fitzgerald PM, McKeever BM, Leu CT, Heimbach JC, Herber WK, Sigal IS, Darke PL, Springer JP (1989) Three-dimensional structure of aspartyl protease from human immunodeficiency virus HIV-1. *Nature* 337: 615–620

20 McQuade TJ, Tomasselli AG, Liu L, Karacostas V, Moss B, Sawyer TK, Heinrikson RL, Tarpley WG (1990) A synthetic HIV-1 protease inhibitor with antiviral activity arrests HIV-like particle maturation. *Science* 247: 454–456

21 Ashorn P, McQuade TJ, Thaisrivongs S, Tomasselli AG, Tarpley WG, Moss B (1990) An inhibitor of the protease blocks maturation of human and simian immunodeficiency viruses and spread of infection. *Proc Natl Acad Sci USA* 87: 7472–7476

22 Roberts NA, Martin JA, Kinchington D, Broadhurst AV, Craig JC, Duncan IB, Galpin SA, Handa

BK, Kay J, Krohn A et al. (1990) Rational design of peptide-based HIV proteinase inhibitors. *Science* 248: 358–361

23 Craig JC, Duncan IB, Hockley D, Grief C, Roberts NA, Mills JS (1991) Antiviral properties of Ro 31-8959, an inhibitor of human immunodeficiency virus (HIV) proteinase. *Antiviral Res* 16: 295–305

24 Vacca JP, Dorsey BD, Schleif WA, Levin RB, McDaniel SL, Darke PL, Zugay J, Quintero JC, Blahy OM, Roth E et al. (1994) L-735,524: an orally bioavailable human immunodeficiency virus type 1 protease inhibitor. *Proc Natl Acad Sci USA* 91: 4096–4100

25 Kempf D, Marsh KC, Denissen JF, McDonald E, Vasavanonda S, Flentge CA, Green BE, Fino L, Park CH, Kong XP et al. (1995) ABT-538 is a potent inhibitor of human immunodeficiency virus protease and has high oral bioavailability in humans. *Proc Natl Acad Sci USA* 92: 2484–2488

26 Markowitz M, Saag M, Powderly WG, Hurley AM, Hsu A, Valdes JM, Henry D, Sattler F, La Marca A, Leonard JM et al. (1995) A preliminary study of ritonavir, an inhibitor of HIV-1 protease, to treat HIV-1 infection. *N Engl J Med* 333: 1534–1539

27 Collier AC, Coombs RW, Schoenfeld DA, Bassett RL, Timpone J, Baruch A, Jones M, Facey K, Whitacre C, McAuliffe VJ et al. (1996) Treatment of human immunodeficiency virus infection with saquinavir, zidovudine, and zalcitabine. AIDS Clinical Trials Group. *N Engl J Med* 334: 1011–1017

28 Gulick RM, Mellors JW, Havlir D, Eron JJ, Gonzalez C, McMahon D, Richman DD, Valentine FT, Jonas L, Meibohm A et al. (1997) Treatment with indinavir, zidovudine, and lamivudine in adults with human immunodeficiency virus infection and prior antiretroviral therapy. *N Engl J Med* 337: 734–739

29 Hammer SM, Squires KE, Hughes MD, Grimes JM, Demeter LM, Currier JS, Eron JJ Jr, Feinberg JE, Balfour HH Jr, Deyton LR et al. (1997) A controlled trial of two nucleoside analogues plus indinavir in persons with human immunodeficiency virus infection and CD4 cell counts of 200 per cubic millimeter or less. *N Engl J Med* 337: 725–733

30 Cameron DW, Heath-Chiozzi M, Danner S, Cohen C, Kravcik S, Maurath C, Sun E, Henry D, Rode R, Potthoff A et al. (1998) Randomised placebo-controlled trial of ritonavir in advanced HIV-1 disease. *Lancet* 351: 543–549

31 Mouton Y, Alfandari S, Valette M, Cartier F, Dellamonica P, Humbert G, Lang JM, Massip P, Mechali D, Leclercq P et al. (1997) Impact of protease inhibitors on AIDS-defining events and hospitalizations in 10 French AIDS reference centres. *AIDS* 11: F101–F105

32 Palella FJ Jr, Delaney KM, Moorman AC, Loveless MO, Fuhrer J, Satten GA, Aschman DJ, Holmberg SD (1998) Declining morbidity and mortality among patients with advanced human immunodeficiency virus infection. HIV Outpatient Study Investigators. *N Engl J Med* 338: 853–860

33 Richman DD, Morton SC, Wrin T, Hellmann N, Berry S, Shapiro MF, Bozzette SA (2004) The prevalence of antiretroviral drug resistance in the United States. *AIDS* 18: 1393–1401

34 Klatzmann D, Champagne E, Chamaret S, Gruest J, Guetard D, Hercend T, Gluckman JC, Montagnier L (1984) T-lymphocyte T4 molecule behaves as the receptor for human retrovirus LAV. *Nature* 312: 767–768

35 Dalgleish AG, Beverley PC, Clapham PR, Crawford DH, Greaves MF, Weiss RA (1984) The CD4 (T4) antigen is an essential component of the receptor for the AIDS retrovirus. *Nature* 312: 763–767

36 McDougal JS, Kennedy MS, Sligh JM, Cort SP, Mawle A, Nicholson JK (1986) Binding of HTLV-III/LAV to T4^{+} T cells by a complex of the 110K viral protein and the T4 molecule. *Science* 231: 382–385

37 Smith DH, Byrn RA, Marsters SA, Gregory T, Groopman JE, Capon DJ (1987) Blocking of HIV-1 infectivity by a soluble, secreted form of the CD4 antigen. *Science* 238: 1704–1707

38 Fisher RA, Bertonis JM, Meier W, Johnson VA, Costopoulos DS, Liu T, Tizard R, Walker BD, Hirsch MS, Schooley RT et al. (1988) HIV infection is blocked *in vitro* by recombinant soluble CD4. *Nature* 331: 76–78

39 Hussey RE, Richardson NE, Kowalski M, Brown NR, Chang HC, Siliciano RF, Dorfman T, Walker B, Sodroski J, Reinherz EL (1988) A soluble CD4 protein selectively inhibits HIV replication and syncytium formation. *Nature* 331: 78–81

40 Deen KC, McDougal JS, Inacker R, Folena-Wasserman G, Arthos J, Rosenberg J, Maddon PJ, Axel R, Sweet RW (1988) A soluble form of CD4 (T4) protein inhibits AIDS virus infection. *Nature* 331: 82–84

41 Traunecker A, Luke W, Karjalainen K (1988) Soluble CD4 molecules neutralize human immunodeficiency virus type 1. *Nature* 331: 84–86
42 Schooley RT, Merigan TC, Gaut P, Hirsch MS, Holodniy M, Flynn T, Liu S, Byington RE, Henochowicz S, Gubish E et al. (1990) Recombinant soluble CD4 therapy in patients with the acquired immunodeficiency syndrome (AIDS) and AIDS-related complex. A phase I-II escalating dosage trial. *Ann Intern Med* 112: 247–253
43 Kahn JO, Allan JD, Hodges TL, Kaplan LD, Arri CJ, Fitch HF, Izu AE, Mordenti J, Sherwin JE, Groopman JE et al. (1990) The safety and pharmacokinetics of recombinant soluble CD4 (rCD4) in subjects with the acquired immunodeficiency syndrome (AIDS) and AIDS-related complex. *Ann Intern Med* 112: 254–261
44 Schacker T, Coombs RW, Collier AC, Zeh JE, Fox I, Alam J, Nelson K, Eggert E, Corey L (1994) The effects of high-dose recombinant soluble CD4 on human immunodeficiency virus type 1 viremia. *J Infect Dis* 169: 37–40
45 Schacker T, Collier AC, Coombs R, Unadkat JD, Fox I, Alam J, Wang JP, Eggert E, Corey L (1995) Phase I study of high-dose, intravenous rsCD4 in subjects with advanced HIV-1 infection. *J Acquir Immune Defic Syndr Hum Retrovirol* 9: 145–152
46 Daar ES, Li XL, Moudgil T, Ho DD (1990) High concentrations of recombinant soluble CD4 are required to neutralize primary human immunodeficiency virus type 1 isolates. *Proc Natl Acad Sci USA* 87: 6574–6578
47 Koot M, Keet IP, Vos AH, de Goede RE, Roos MT, Coutinho RA, Miedema F, Schellekens PT, Tersmette M (1993) Prognostic value of HIV-1 syncytium-inducing phenotype for rate of $CD4^+$ cell depletion and progression to AIDS. *Ann Intern Med* 118, 681–688
48 Cocchi F, DeVico AL, Garzino-Demo A, Arya SK, Gallo RC, Lusso P (1995) Identification of RANTES, MIP-1 alpha, and MIP-1 beta as the major HIV-suppressive factors produced by $CD8^+$ T cells. *Science* 270: 1811–1815
49 Dragic T, Litwin V, Allaway GP, Martin SR, Huang Y, Nagashima KA, Cayanan C, Maddon PJ, Koup RA, Moore JP et al. (1996) HIV-1 entry into $CD4^+$ cells is mediated by the chemokine receptor CC-CKR-5. *Nature* 381: 667–673
50 Deng H, Liu R, Ellmeier W, Choe S, Unutmaz D, Burkhart M, Di Marzio P, Marmon S, Sutton RE, Hill CM et al. (1996) Identification of a major co-receptor for primary isolates of HIV-1. *Nature* 381: 661–666
51 Feng Y, Broder CC, Kennedy PE, Berger EA (1996) HIV-1 entry cofactor: functional cDNA cloning of a seven-transmembrane, G protein-coupled receptor. *Science* 272: 872–877
52 Choe H, Farzan M, Sun Y, Sullivan N, Rollins B, Ponath PD, Wu L, Mackay CR, LaRosa G, Newman W et al. (1996) The beta-chemokine receptors CCR3 and CCR5 facilitate infection by primary HIV-1 isolates. *Cell* 85: 1135–1148
53 Doranz BJ, Rucker J, Yi Y, Smyth RJ, Samson M, Peiper SC, Parmentier M, Collman RG, Doms RW (1996) A dual-tropic primary HIV-1 isolate that uses fusin and the beta-chemokine receptors CKR-5, CKR-3, and CKR-2b as fusion cofactors. *Cell* 85: 1149–1158
54 Strizki JM, Xu S, Wagner NE, Wojcik L, Liu J, Hou Y, Endres M, Palani A, Shapiro S, Clader JW et al. (2001) SCH-C (SCH 351125), an orally bioavailable, small molecule antagonist of the chemokine receptor CCR5, is a potent inhibitor of HIV-1 infection *in vitro* and *in vivo*. *Proc Natl Acad Sci USA* 98: 12718–12723
55 Reynes J, Rouzier R, Kanouni T, Kanouni T, Baillat V, Baroudy B, Keung A, Hogan C, Markowitz M, Laughlin M (2002) SCH C: Safety and antiviral effects of a CCR5 receptor antagonist in HIV-1 infected subjects. 9th Conference on Retroviruses and Opportunistic Infections 2002, Seattle, WA, February 24–28, 2002, abstract no. 1
56 Strizki JM, Tremblay C, Xu S, Wojcik L, Wagner N, Gonsiorek W, Hipkin RW, Chou CC, Pugliese-Sivo C, Xiao Y et al. (2005) Discovery and characterization of vicriviroc (SCH 417690), a CCR5 antagonist with potent activity against human immunodeficiency virus type 1. *Antimicrob Agents Chemother* 49: 4911–4919
57 De Clercq E, Yamamoto N, Pauwels R, Baba M, Schols D, Nakashima H, Balzarini J, Debyser Z, Murrer BA, Schwartz D et al. (1992) Potent and selective inhibition of human immunodeficiency virus (HIV)-1 and HIV-2 replication by a class of bicyclams interacting with a viral uncoating event. *Proc Natl Acad Sci USA*. 89: 5286–5290
58 De Clercq E, Yamamoto N, Pauwels R, Balzarini J, Witvrouw M, De Vreese K, Debyser Z, Rosenwirth B, Peichl P, Datema R et al. (1994) Highly potent and selective inhibition of human immunodeficiency virus by the bicyclam derivative JM3100. *Antimicrob Agents Chemother* 38:

668–674
59 Schols D, Este JA, Henson G, De Clercq E (1997) Bicyclams, a class of potent anti-HIV agents, are targeted at the HIV coreceptor fusin/CXCR-4. *Antiviral Res* 35: 147–156
60 Donzella GA, Schols D, Lin SW, Este JA, Nagashima KA, Maddon PJ, Allaway GP, Sakmar TP, Henson G, De Clercq E, Moore JP (1998) AMD3100, a small molecule inhibitor of HIV-1 entry via the CXCR4 co-receptor. *Nat Med* 4: 72–77
61 Hendrix CW, Flexner C, MacFarland RT, Giandomenico C, Fuchs EJ, Redpath E, Bridger G, Henson GW (2000) Pharmacokinetics and safety of AMD-3100, a novel antagonist of the CXCR-4 chemokine receptor, in human volunteers. *Antimicrob Agents Chemother* 44: 1667–1673
62 Hendrix CW, Collier AC, Lederman MM, Schols D, Pollard RB, Brown S, Jackson JB, Coombs RW, Glesby MJ, Flexner CW et al. (2004) Safety, pharmacokinetics, and antiviral activity of AMD3100, a selective CXCR4 receptor inhibitor, in HIV-1 infection. *J Acquir Immune Defic Syndr* 37: 1253–1262
63 Gallaher WR, Ball JM, Garry RF, Griffin MC, Montelaro RC (1989) A general model for the transmembrane proteins of HIV and other retroviruses. *AIDS Res Hum Retroviruses* 5: 431–440
64 Delwart EL, Mosialos G, Gilmore T (1990) Retroviral envelope glycoproteins contain a "leucine zipper"-like repeat. *AIDS Res Hum Retroviruses* 6: 703–706
65 Jiang S, Lin K, Strick N, Neurath AR (1993) HIV-1 inhibition by a peptide. *Nature* 365: 113
66 Chen CH, Matthews TJ, McDanal CB, Bolognesi DP, Greenberg ML (1995) A molecular clasp in the human immunodeficiency virus (HIV) type 1 TM protein determines the anti-HIV activity of gp41 derivatives: implication for viral fusion. *J Virol* 69: 3771–3777
67 Wild C, Greenwell T, Matthews T (1993) A synthetic peptide from HIV-1 gp41 is a potent inhibitor of virus-mediated cell-cell fusion. *AIDS Res Hum Retroviruses* 9: 1051–1053
68 Kilby JM, Hopkins S, Venetta TM, DiMassimo B, Cloud GA, Lee JY, Alldredge L, Hunter E, Lambert D, Bolognesi D et al. (1998) Potent suppression of HIV-1 replication in humans by T-20, a peptide inhibitor of gp41-mediated virus entry. *Nat Med* 4: 1302–1307
69 Lalezari JP, Henry K, O'Hearn M, Montaner JS, Piliero PJ, Trottier B, Walmsley S, Cohen C, Kuritzkes DR, Eron JJ Jr et al. (2003) Enfuvirtide, an HIV-1 fusion inhibitor, for drug-resistant HIV infection in North and South America. *N Engl J Med* 348: 2175–2185
70 Lazzarin A, Clotet B, Cooper D, Reynes J, Arasteh K, Nelson M, Katlama C, Stellbrink HJ, Delfraissy JF, Lange J et al. (2003) Efficacy of enfuvirtide in patients infected with drug-resistant HIV-1 in Europe and Australia. *N Engl J Med* 348: 2186–2195
71 Panel on Clinical Practices for the Treatment of HIV Infection (2006) Guidelines for the use of antiretroviral agents in HIV-infected adults and adolescents. Available at <www.aids info.org>. October 10, 2006 update
72 Hammer SM, Saag MS, Schechter M, Montaner JS, Schooley RT, Jacobsen DM, Thompson MA, Carpenter CC, Fischl MA, Gazzard BG et al. (2006) Treatment for adult HIV infection: 2006 recommendations of the International AIDS Society-USA panel. *JAMA* 296: 827–843
73 Trimeris, Fuzeon Package Insert (April 2005)
74 Allaway GP, Davis-Bruno KL, Beaudry GA, Garcia EB, Wong EL, Ryder AM, Hasel KW, Gauduin MC, Koup RA, McDougal JS et al. (1995) Expression and characterization of CD4-IgG2, a novel heterotetramer that neutralizes primary HIV type 1 isolates. *AIDS Res Hum Retroviruses* 11: 533–539
75 Jacobson JM, Lowy I, Fletcher CV, O'Neill TJ, Tran DN, Ketas TJ, Trkola A, Klotman ME, Maddon PJ, Olson WC et al. (2000) Single-dose safety, pharmacology, and antiviral activity of the human immunodeficiency virus (HIV) type 1 entry inhibitor PRO 542 in HIV-infected adults. *J Infect Dis* 182: 326–329
76 Jacobson JM, Israel RJ, Lowy I, Ostrow NA, Vassilatos LS, Barish M, Tran DN, Sullivan BM, Ketas TJ, O'Neill TJ et al. (2004) Treatment of advanced human immunodeficiency virus type 1 disease with the viral entry inhibitor PRO 542. *Antimicrob Agents Chemother* 48: 423–429
77 Shearer WT, Israel RJ, Starr S, Fletcher CV, Wara D, Rathore M, Church J, DeVille J, Fenton T, Graham B et al. (2000) Recombinant CD4-IgG2 in human immunodeficiency virus type 1-infected children: phase 1/2 study. *J Infect Dis* 182: 1774–1779
78 Burkly LC, Olson D, Shapiro R, Winkler G, Rosa JJ, Thomas DW, Williams C, Chisholm P (1992) Inhibition of HIV infection by a novel CD4 domain 2-specific monoclonal antibody. Dissecting the basis for its inhibitory effect on HIV-induced cell fusion. *J Immunol* 149: 1779–1787
79 Kuritzkes DR, Jacobson J, Powderly WG, Godofsky E, DeJesus E, Haas F, Reimann KA, Larson JL, Yarbough PO, Curt V et al. (2004) Antiretroviral activity of the anti-CD4 monoclonal antibody

TNX-355 in patients infected with HIV type 1. *J Infect Dis* 189: 286–291
80 Jacobson JM, Kuritzkes DR, Godofsky E, DeJesus E, Lewis S, Jackson J, Frazier K, Fagan EA, Shanahan WR (2004) Phase 1b study of the anti-CD4 monoclonal antibody TNX-355 in HIV-1-infected subjects: safety and antiretroviral activity of multiple doses. In: abstracts of the 11th Conference on Retroviruses and Opportunistic Infections, San Francisco, CA, February 8–11, 2004, abstract no. 536
81 Norris D, Moreales J, Godofsky E, Garcia F, Hardwicke R, Lewis S (2006) TNX-355, in combination with optimized background regimen (OBR), achieves statistically significant viral load reduction and CD4 cell count increase when compared with OBR alone in phase 2 study at 48 weeks. In: Abstracts of the XVI International AIDS Conference, Toronto, Canada, August 13–18, 2006, abstract no. THLB0218
82 Lin PF, Blair W, Wang T, Spicer T, Guo Q, Zhou N, Gong YF, Wang HG, Rose R, Yamanaka G et al. (2003) A small molecule HIV-1 inhibitor that targets the HIV-1 envelope and inhibits CD4 receptor binding. *Proc Natl Acad Sci USA* 100: 11013–11018
83 Guo Q, Ho HT, Dicker I, Fan L, Zhou N, Friborg J, Wang T, McAuliffe BV, Wang HG, Rose RE et al. (2003) Biochemical and genetic characterizations of a novel human immunodeficiency virus type 1 inhibitor that blocks gp120-CD4 interactions. *J Virol* 77: 10528–10536
84 Hanna G, Yan J-H, Fiske W, Materson T, Zhang D, Grasela D (2004) Safety, tolerability, and pharmacokinetics of a novel small-molecule HIV-1 attachment inhibitor, BMS-488043, after single and multiple oral doses in healthy subjects. In: Abstracts of the 11th Conference on Retroviruses and Opportunistic Infections, San Francisco, CA, February 8–11, 2004, abstract no. 535
85 Hanna G, Lalezari J, Hellinger J, Wohl D, Masterson T, Fiske W, Kadow J, Line P, Giordano M, Colonno R, Grasela D (2004) Antiviral activity, safety, and tolerability of a novel, oral small-molecule HIV-1 attachment inhibitor, BMS-488043, in HIV-1-infected subjects. In: Abstracts of the 11th Conference on Retroviruses and Opportunistic Infections, San Francisco, CA, February 8–11, 2004, abstract no. 141
86 Watson C, Jenkinson S, Kazmierski W, Kenakin T (2005) The CCR5 receptor-based mechanism of action of 873140, a potent allosteric noncompetitive HIV entry inhibitor. *Mol Pharmacol* 67: 1268–1282
87 Demarest J, Shibayama S, Ferris R, Vavro C, St Clair M, Boone L (2004) A novel CCR5 antagonist, 873140, exhibits potent *in vitro* anti-HIV activity. In: Abstracts of the XV International AIDS Conference, Bangkok, Thailand, July 11–16, 2004, abstract no. WeOrA1231
88 Adkison KK, Shachoy-Clark A, Fang L, Lou Y, O'Mara K, Berrey MM, Piscitelli SC (2005) Pharmacokinetics and short-term safety of 873140, a novel CCR5 antagonist, in healthy adult subjects. *Antimicrob Agents Chemother* 49: 2802–2806
89 Lalezari J, Thompson M, Kumar P, Piliero P, Davey R, Patterson K, Shachoy-Clark A, Adkison K, Demarest J, Lou Y et al. (2005) Antiviral activity and safety of 873140, a novel CCR5 antagonist, during short-term monotherapy in HIV-infected adults. *AIDS* 19: 1443–1448
90 GlaxoSmithKline Press Release (October 25, 2005)
91 Dorr P, Westby M, Dobbs S, Griffin P, Irvine B, Macartney M, Mori J, Rickett G, Smith-Burchnell C, Napier C et al. (2005) Maraviroc (UK-427,857), a potent, orally bioavailable, and selective small-molecule inhibitor of chemokine receptor CCR5 with broad-spectrum anti-human immunodeficiency virus type 1 activity. *Antimicrob Agents Chemother* 49: 4721–4732
92 Price DA, Armour D, de Groot M, Leishman D, Napier C, Perros M, Stammen BL, Wood A (2006) Overcoming HERG affinity in the discovery of the CCR5 antagonist maraviroc. *Bioorg Med Chem Lett* 16: 4633–4637
93 Abel S, Russel D, Ridgway C, Muirhead G (2005) Overivew of the drug-drug interaction data for maraviroc (MVC, UK-427,857). In: Abstracts of the 6th International Workshop on Clinical Pharmacology of HIV Therapy, Quebec, Canada, April 28–30, 2005, abstract no. 76
94 Fatkenheuer G, Pozniak AL, Johnson MA, Plettenberg A, Staszewski S, Hoepelman AI, Saag MS, Goebel FD, Rockstroh JK, Dezube BJ et al. (2005) Efficacy of short-term monotherapy with maraviroc, a new CCR5 antagonist, in patients infected with HIV-1. *Nat Med* 11: 1170–1172
95 Mayer H (2006) Forum Working Group on Chemokine Antagonists in HIV Therapy: RoundTable 3: Focus on Lymphomas and Malignancies, Washington, DC, May 30, 2006
96 Pfizer Press Release (January 24, 2006)
97 Schurmann D, Rouzier R, Nougarede R, Reynes J, Fatkenheuer G, Raffi F, Michelet C, Tarral A, Hoffmann C, Kiunke J et al. (2004) SCH D: Antiviral activity of a CCR5 receptor antagonist. In: Abstracts for the 11th Conference on Retroviruses and Opportunistic Infections, San Francisco,

CA, February 8–11, 2004, abstract no. 140LB
98 Seiberling M, Kraan M, Keung A, Martinho M, Sansone A (2005) Similar increase in SCH 417690 plasma exposure with coadministration of varying doses of ritonavir in healthy volunteers. In: Abstracts of the 3rd International AIDS Society Conference on Pathogenesis and Treatment, July 24–27, 2005, Rio de Janeiro, Brazil, abstract no. TuPe3.1B06
99 Greaves W, Landovitz R, Fatkenheuer G, Hoffmann C, Antunes F, Angel J, Boparai N, Knepp D, Keung A, Dunkle L (2006) Late virologic breakthrough in treatment-naïve patients on a regimen of Combivir + vicriviroc. In: Abstracts of the 13th Conference on Retroviruses and Opportunistic Infections, February 5–8, 2006, Denver, CO, abstract no. 161LB
100 Gulick R, Su Z, Flexner C, Hughes M, Skolnik P, Wilkin T, Gross R, Krambrink A, Coakley E, Greaves WL et al (2007) Phase II study of the safety and efficacy of vicriviroc, a CCR inhibitor in HIV-infected treatment experienced patients: ACTG 52. *J Infect Dis* (in press)
101 Martinson JJ, Chapman NH, Rees DC, Liu YT, Clegg JB. (1997) Global distribution of the CCR5 gene 32-basepair deletion. *Nat Genet* 16: 100–103
102 Samson M, Libert F, Doranz BJ, Rucker J, Liesnard C, Farber CM, Saragosti S, Lapoumeroulie C, Cognaux J, Forceille C et al. (1996) Resistance to HIV-1 infection in caucasian individuals bearing mutant alleles of the CCR-5 chemokine receptor gene. *Nature* 382: 722–725
103 Garred P, Madsen HO, Petersen J, Marquart H, Hansen TM, Freiesleben Sorensen S, Volck B, Svejgaard A, Andersen V. (1998) CC chemokine receptor 5 polymorphism in rheumatoid arthritis. *J Rheumatol* 25: 1462–1465
104 Hellier S, Frodsham AJ, Hennig BJ, Klenerman P, Knapp S, Ramaley P, Satsangi J, Wright M, Zhang L, Thomas HC et al. (2003) Association of genetic variants of the chemokine receptor CCR5 and its ligands, RANTES and MCP-2, with outcome of HCV infection. *Hepatology* 38: 1468–1476
105 Fischereder M, Luckow B, Hocher B, Wuthrich RP, Rothenpieler U, Schneeberger H, Panzer U, Stahl RA, Hauser IA, Budde K et al. (2001) CC chemokine receptor 5 and renal-transplant survival. *Lancet* 357: 1758–1761
106 Moench C, Uhrig A, Lohse AW, Otto G. (2004) CC chemokine receptor 5delta32 polymorphism-a risk factor for ischemic-type biliary lesions following orthotopic liver transplantation. *Liver Transpl* 10: 434–439
107 Glass WG, McDermott DH, Lim JK, Lekhong S, Yu SF, Frank WA, Pape J, Cheshier RC, Murphy PM. (2006) CCR5 deficiency increases risk of symptomatic West Nile virus infection. *J Exp Med* 203: 35–40
108 Lalezari JP, Thompson M, Kumar P, Piliero P, Davey R, Patterson K, Shachoy-Clark A, Atkison K, Demarest J, Lou Y, et al (2005) Antiviral activity and safety of 873140, a novel CCR5 antagonist, during short-term monotherapy in HIV-infected adults. *AIDS* 19: 1443–1448
109 Westby M, Lewis M, Whitcomb J, Youle M, Pozniak AL, James IT, Jenkins TM, Perros M, van der Ryst E. (2006) Emergence of CXCR4-using human immunodeficiency virus type 1 (HIV-1) variants in a minority of HIV-1-infected patients following treatment with the CCR5 antagonist maraviroc is from a pretreatment CXCR4-using virus reservoir. *J Virol* 80: 4909–4920
110 Mayer H, van der Ryst E, Saag M, Clotet B, Fatkenheuer G, Clumeck N, Turner K, Goodrich JM. (2006) Safety and efficacy of maraviroc (MVC), a novel CCR5 antagonist, when used in combination with optimized background therapy (OBT) for the treatment of antiretroviral-experienced subjects infected with dual/mixed tropic HIV-1: 24-week results of a phase 2b exploratory trial. IN: Abstracts of the XVI International AIDS Conference, Toronto, Canada, August 13–18, 2006, abstract no. THLB0215
111 Schols D, Claes S, Hatse S, Princen K, Vermeire K, De Clercq E, Skerlj R, Bridger G, Calandra G. (2003) Anti-HIV profile of AMD070, an orally bioavailble CXCR4 antagonist H. IN: Abstracts of the 10th Conference on Retroviruses and Opportunistic Infections, Boston, MA, February 10–14, 2003, abstract no. 563
112 Stone N, Dunaway S, Flexner C, Calandra G, Wiggins I, Conley J, Snyder S, Tierney C, Hendrix CW. (2004) Biologic activity of an orally bioavailable CXCR4 antagonist in human subjects. IN: Abstracts of the XV International AIDS Conference, Bangkok, Thailand, July 11–16, 2004, abstract no. TuPeB4475
113 Anormed Press Release (March 17, 2006)
114 Mink M, Mosier SM, Janumpalli S, Davison D, Jin L, Melby T, Sista P, Erickson J, Lambert D, Stanfield-Oakley SA et al. (2005) Impact of human immunodeficiency virus type 1 gp41 amino acid substitutions selected during enfuvirtide treatment on gp41 binding and antiviral potency of

enfuvirtide *in vitro*. *J Virol* 79: 12447–12454

115 Eron JJ, Gulick RM, Bartlett JA, Merigan T, Arduino R, Kilby JM, Yangco B, Diers A, Drobnes C, DeMasi R et al. (2004) Short-term safety and antiretroviral activity of T-1249, a second-generation fusion inhibitor of HIV. *J Infect Dis* 189: 1075–1083

116 Dai SJ, Dou GF, Qiang XH, Song HF, Tang ZM, Liu DS, Liu XW, Yang LM, Zheng YT, Liang Q (2005) Pharmacokinetics of sifuvirtide, a novel anti-HIV-1 peptide, in monkeys and its inhibitory concentration *in vitro*. *Acta Pharmacol Sin* 26: 1274–1280

117 Delmedico M, Bray B, Cammack N, Davison D, Dwyer J, Frick L, Tvermoes N, Wring S, Zhang H, Greenberg M. (2006) Next generation HIV peptide fusion inhibitor candidates achieve potent, durable suppression of virus replication *in vitro* and improved pharmacokinetic properties. In: Abstracts of the 13th Conference on Retroviruses and Opportunistic Infections, February 5–8, 2006, Denver, CO, abstract no. 48

118 Veazey RS, Springer MS, Marx PA, Dufour J, Klasse PJ, Moore JP. (2005) Protection of macaques from vaginal SHIV challenge by an orally delivered CCR5 inhibitor. *Nat Med* 11: 1293–1294

119 Grinsztejn B, Nguyen BY, Katlama C, Gatell J, Lazzarin A, Vittecoq D, Gonzalez C, Chen J, Isaacs R (2006) Potent antiretroviral effect of MK-0518, a novel HIV-1 integrase inhibitor, in patients with triple-class resistant virus. In: Abstracts of the 13th Conference on Retroviruses and Opportunistic Infections, February 5–8, 2006, Denver, CO, abstract no. 159LB

120 DeJesus E, Berger D, Markowitz M, Cohen C, Hawkins T, Ruane P, Elion R, Farthing C, Zhong L, Cheng AK et al. (2006) Antiviral activity, pharmacokinetics, and dose response of the HIV-1 integrase inhibitor GS-9137 (JTK-303) in treatment-naive and treatment-experienced patients. *J Acquir Immune Defic Syndr* 43: 1–5

121 Smith P, Forrest A, Beatty G, Jacobson J, Lalezari J, Eron J, Pollard R, Saag M, Doto J, Martin D (2006) Pharmacokinetics/pharmacodynamics of PA-457 in a 10-day multiple dose monotherapy trial in HIV-infected patients. In: Abstracts of the 13th Conference on Retroviruses and Opportunistic Infections, February 5–8, 2006, Denver, CO, abstract no. 52

122 Tremblay CL, Kollmann C, Giguel F, Chou TC, Hirsch MS. (2000) Strong *in vitro* synergy between the fusion inhibitor T-20 and the CXCR4 blocker AMD-3100. *J Acquir Immune Defic Syndr* 25: 99–102

123 Tremblay CL, Giguel F, Kollmann C, Guan Y, Chou TC, Baroudy BM, Hirsch MS. (2002) Anti-human immunodeficiency virus interactions of SCH-C (SCH 351125), a CCR5 antagonist, with other antiretroviral agents *in vitro*. *Antimicrob Agents Chemother* 46: 1336–1339

124 Reeves JD, Gallo SA, Ahmad N, Miamidian JL, Harvey PE, Sharron M, Pohlmann S, Sfakianos JN, Derdeyn CA, Blumenthal R et al. (2002) Sensitivity of HIV-1 to entry inhibitors correlates with envelope/coreceptor affinity, receptor density, and fusion kinetics. *Proc Natl Acad Sci USA* 99: 16249–16254

125 Schols D, Vermeire K, Hatse S, Princen K, De Clercq E, Calandra G, Ricker S, Nelson K, Labrecque J, Bogucki D et al. (2004) *In vitro* anti-HIV activity profile of AMD887, a novel CCR5 antagonist, in combination with the CXCR4 inhibitor AMD 070. In: Abstracts of the 11th Conference on Retroviruses and Opportunistic Infections, San Francisco, CA, February 8–11, 2004, abstract no. 539

126 Moore JP, Kitchen SG, Pugach P, Zack JA. (2004) The CCR5 and CXCR4 coreceptors – central to understanding the transmission and pathogenesis of human immunodeficiency virus type 1 infection. *AIDS Res Hum Retroviruses* 20: 111–126

127 Sista PR, Melby T, Davison D, Jin L, Mosier S, Mink M, Nelson EL, DeMasi R, Cammack N, Salgo MP et al. (2004) Characterization of determinants of genotypic and phenotypic resistance to enfuvirtide in baseline and on-treatment HIV-1 isolates. *AIDS* 18: 1787–1794

Index

The MDT-Series
Milestones in Drug Therapy

The discovery of drugs is still an unpredictable process. Breakthroughs are often the result of a combination of factors, including serendipidity, rational strategies and a few individuals with novel ideas. *Milestones in Drug Therapy* highlights new therapeutic developments that have provided significant steps forward in the fight against disease. Each book deals with an individual drug or drug class that has altered the approach to therapy. Emphasis is placed on the scientific background to the discoveries and the development of the therapy, with an overview of the current state of knowledge provided by experts in the field, revealing also the personal stories behind these milestone developments. The series is aimed at a broad readership, covering biotechnology, biochemistry, pharmacology and clinical therapy.

Forthcoming titles:

Treatment of Psoriasis, J.M. Weinberg (Editor), 2007
Tamoxifen and Beyond, V.C. Jordan (Author), 2007
Pharmacotherapy of Obesity, J.P.H. Wilding (Editor), 2007

Published volumes:

Drugs Affecting Growth of Tumours, H.M. Pinedo, C.H. Smorenburg (Editors), 2006
TNF-alpha Inhibitors, J.M. Weinberg, R. Buchholz (Editor), 2006
Aromatase Inhibitors, B.J.A. Furr (Editor), 2006
Cannabinoids as Therapeutics, R. Mechoulam (Editor), 2005
St. John's Wort and its Active Principles in Anxiety and Depression, W.E. Müller (Editor), 2005
Drugs for Relapse Prevention of Alcoholism, R. Spanagel, K. Mann (Editors), 2005
COX-2 Inhibitors, M. Pairet, J. Van Ryn (Editors), 2004
Calcium Channel Blockers, T. Godfraind (Author), 2004
Sildenafil, U. Dunzendorfer (Editor), 2004
Hepatitis Prevention and Treatment, J. Colacino, B.A. Heinz (Editors), 2004
Combination Therapy of AIDS, E. De Clercq, A.M. Vandamme (Editors), 2004
Cognitive Enhancing Drugs, J. Buccafusco (Editor), 2004
Fluoroquinolone Antibiotics, A.R. Ronald, D. Low (Editors), 2003
Erythropoietins and Erythropoiesis, G. Molineux, M. Foote, S. Elliott (Editors), 2003
Macrolide Antibiotics, W. Schönfeld, H. Kirst (Editors), 2002

HMG CoA Reduktase Inhibitors, G. Schmitz, M. Torzewski (Editors), 2002
Antidepressants, B.E. Leonard (Editor), 2001
Recombinant Protein Drugs, P. Buckel (Editor), 2001
Glucocorticoids, N. Goulding, R.J. Flower (Editors), 2001
Modern Immunosuppressives, H.-J. Schuurman (Editor), 2001
ACE Inhibitors, P. D'Orleans-Juste, G. Plante (Editors), 2001
Atypical Antipsychotics, A.R. Cools, B.A. Ellenbroek (Editors), 2000
Methotrexate, B.N. Cronstein, J.R. Bertino (Editors), 2000
Anxiolytics, M. Briley, D. Nutt (Editors), 2000
Proton Pump Inhibitors, L. Olbe (Editor), 1999
Valproate, W. Löscher (Editor), 1999